'This fascinating and engaging book challenges us all to make better lives for animals.'
Chris Packham, broadcaster and author of *Back to Nature*

'Marvellous: clear-headed, clear-sighted, rigorously unsentimental but compassionate, impeccably informed and researched, and downright wise. A companionable guide through complex and controversial territory. It deserves to be a canonical text in the welfare debate.'
Charles Foster, author of *Cry of the Wild* and *Being a Beast*

'A thoughtful, well-informed contribution to the animal-welfare and conservation debate.'
Jane Dalton, *The Independent*

'Animals are sentient beings, with capacity to experience both suffering and joy. Alick Simmons takes this as his starting point and brings his huge veterinary experience to this book, along with his honesty and desire for reform. The result makes for a highly informative and thought-provoking read.'
Joyce D'Silva, Compassion in World Farming

'Thoughtful, informative and firm, Simmons skilfully leads us through the complex maze of animal welfare issues and brings us to a stark realisation - for all that we have done, we must do better.'
Professor Adam Hart, biologist, broadcaster and author of *Unfit for Purpose*

'Refreshingly, Alick Simmons neither castigates nor judges, but rather leads the reader through these contentious issues with gentle, considered and well-reasoned views that are hard to ignore.'
Ruth Tingay, conservationist and co-director of Wild Justice

'He writes with expertise and knowledge, posing questions rather than passing judgement.'
James Chubb, farmer and conservationist

'Simmons' approach pulls no punches, but it is thoughtful and broad-ranging – and ultimately hopeful: rather than pressing a particular conclusion, he encourages his readers to develop their own ethical framework for considering the consequences of their actions.'
Rosie Woodroffe, Biologist, Zoological Society of London

'Wide-ranging and extensively researched, cogently argued while impressively modest, Alick Simmons' debut is an accomplished, thought-provoking work that asks all the right questions of our relationship with animals while supporting us to provide our own answers.'
James Lowen, author of *Much Ado About Mothing*

'A rigorous, balanced and highly readable examination of the various ways we exploit the animals we live alongside. Full of good story-telling and the distilled wisdom from a distinguished career in the field.'
Ian Carter, author of *Rhythms of Nature and Human, Nature*

'This is a lucid, persuasive and deeply thought-provoking contribution on our relationship with animals. In what can be a highly polarised and contested debate Alick Simmons' view is characterised by clarity, honesty, conscience – and above all carefully argued positions across a wide range of circumstances.'
Steve Ormerod, Professor of Ecology, Cardiff University

'Stimulating, challenging, and important.'
Hugh Warwick, ecologist and author of *Linescapes*

'This is an important and challenging book about how humans treat animals . . . Instead of advocating for a single correct viewpoint, Simmons promotes the importance of developing a rational and coherent personal ethical framework, and applying this evidence-based framework consistently across different animal species and societal contexts.'
Rob Thomas, Senior Lecturer in Zoology, Ecology and Data Analysis at Cardiff University School of Biosciences

Treated Like Animals

Treated Like Animals

Improving the Lives of the Creatures We Own, Eat and Use

Alick Simmons

Pelagic Publishing

Published in 2023 by Pelagic Publishing
20–22 Wenlock Road
London N1 7GU, UK

www.pelagicpublishing.com

Treated Like Animals: Improving the Lives of the Creatures We Own, Eat and Use

A CIP record for this book is available from the British Library

ISBN 978-1-78427-341-5 Pbk
ISBN 978-1-78427-342-2 ePub
ISBN 978-1-78427-343-9 ePDF
ISBN 978-1-78427-420-7 Audio

https://doi.org/10.53061/OHKX6488

Cover design: Edward Bettison

Author photo: Fran Stockwell

Typeset in Adobe Garamond Pro by Deanta Global Publishing Services, Chennai, India

Printed in England by TJ Books

To Mrs Mouse

Contents

Preface   xii

 1 The Exploitation of Animals   1
 2 Why Aren't All Animals Treated the Same Way?   21
 3 The Welfare of Farmed Animals: an Overview   44
 4 Grazing Animals: the Best, and Some of the Worst   57
 5 Pigs, Poultry and the Rest   76
 6 Snares, Guns and Poison: the 'Management' of Wildlife   101
 7 Conservation: Exploitation with Clear Limits?   127
 8 Recreation, Sport and a Little Food   144
 9 Pets: Exploitation Begins at Home   157
10 Animals Used in Research   166
11 A Personal Ethical Framework   181
12 Making Sense of It All   190

Notes   213
Glossary and Abbreviations   238
Further Reading   244
Acknowledgements   245
Index   247
About the Author   254

Preface

If thinking, agonising and prevaricating about a particular subject counts as research, then this book might have had one of the longest periods of research of any book, ever. The exploitation of animals, good or bad, has played on my mind for decades, even before I became a veterinarian in the late 1970s.

I concluded many years ago that, when it comes to our interactions with animals, much of what we do, much of what we tolerate and much of what we enjoy is inhumane. And yet I gained a qualification and have had decades of gainful employment where I have actively facilitated exploitative interactions.

To be a veterinarian is an odd calling. Young, often idealistic, people become undergraduate veterinary students for a number of reasons, but studies show that it stems mainly from a 'love' of animals and a desire to relieve suffering. And why not? However, although the undergraduate courses and employee support have improved over the years, nothing can prepare new graduates for the rude awakening that awaits them in their first few months of practice: the profession is demanding, relentless and often lonely.

It is clear that isolation and lack of support can have a profound effect on an individual's mental health. I was lucky, and although I found the first few months hellish, I coped – somehow. But others are less fortunate: suicide rates among veterinarians are some of the highest of any type of employment.[1] Two of my contemporaries killed themselves within five years of qualifying. Thankfully, there is help available; an excellent organisation now exists to support vets – run by vets, and staffed by trained volunteers (www.vetlife.org.uk).

Perhaps poor mental health and the high suicide risk in veterinarians is linked to a disconnect between the expectation of young graduates and what society expects of them. For decades, the veterinary profession and the society it serves have deluded themselves into thinking that the veterinarian exists solely to minister to sick and injured animals. Sure, a lot of what vets do either prevents or relieves suffering. But this is only a small part of the job. Most vets spend most of their time facilitating society's exploitation of animals. This includes ensuring they grow well so we can eat them, ensuring they recover from going lame so we can ride them, and ensuring they aren't diseased so they don't poison us. The rest involves straightening out the inherited and acquired defects brought about by irresponsible breeding and incompetent care. These are the veterinary services we don't like to talk about.

It was into this profession I was thrust a little over 40 years ago. It soon became obvious to me that I was unsuited to private practice, and that the feeling was mutual. However, I was fortunate to be born into a generation, educated in the 1970s, where curious and ambitious people could forge interesting and exciting careers. So I took advantage of the opportunities. Because of an interest in public health I joined the UK government's veterinary service, and I studied tropical veterinary medicine, animal behaviour and welfare, along the way. I had various and varied jobs which periodically required travel around the world on behalf of the UK government and the EU, including extended stays in Belize and Australia. I left public service in 2015, having been the UK government's Deputy Chief Veterinary Officer for the previous eight years. Even with the benefit of hindsight, I remain grateful for the many opportunities and for the challenging times.

Despite being intermittently embroiled in gruelling work on BSE, foot-and-mouth disease and other diseases of farmed animals, I managed to maintain a personal and professional interest in animal welfare; this sat alongside my lifelong passion for wildlife. And since leaving public service, I've managed to join together those twin interests – I now advise two conservation NGOs on wildlife welfare and ethics. Calling it a second career would be a stretch but for me these are small successes of which I am very proud.

It would be an exaggeration to claim that my veterinary career was simply research for this book, but there is no doubt that my experience has inspired a great deal of the content. While we all, with few exceptions, exploit animals, my career has brought me closer than most to almost every type of exploitation. You cannot help be affected by it.

This is a not a book about the veterinary profession, however. It is about animals, how we treat them and how we could do better. Vets have a role to play here, but this is an issue for everyone – for individuals and for society as a whole. I haven't played down my career and my profession but, deliberately, the focus is on the animals we exploit.

I am the product of my upbringing and the child of parents who suffered the privations of the Second World War. They knew the value of food and the importance of protein for growing children. That meant three square meals a day, lots of milk and a roast on Sunday. My late mother would be as likely to consider serving a meat-free meal to my three siblings and me as to make a stew of horse meat. But attitudes and tastes change over time – and, ever curious, in her later years my mother became more interested in the provenance of her food, and in vegetarian cooking. I inherited her curiosity.

I am not vegetarian but I am picky about where my meat comes from and avoid certain types altogether. Periodically I go through several weeks of eating only vegetarian food, but I haven't yet gone the whole hog. My dithering might be explained by my upbringing, my tastes being set by my teens, but I don't think that's the whole story. It is because I believe that limited and highly specific forms of animal exploitation can be justified.

While it is almost certainly true that we should eat a great deal less meat, and that many other forms of exploitation need to be reduced, reformed or prohibited, I cannot yet subscribe to the idea that all animal exploitation must cease. Certain types of animal production are good for the environment, some time-limited interventions are necessary to protect threatened species, safety and livelihoods, and some research using animals cannot be avoided if we are to tackle disease in human society.

But change, perhaps radical change, is necessary. All exploitation can be reduced, replaced and refined. The citizen can be better informed, and can be empowered to exert more influence. Operatives and animal keepers can be held to account more effectively than they are at present.

This book is intended to inform and stimulate curious people, people who are questioning their diet and attitudes to animals and who want more information before making a decision about what they will tolerate.

The nature and extent of animal exploitation is a matter of ethics, individual and societal. Exercising choice on the basis of an ethical position, whether individually or collectively, should be on the basis of evidence. After reading this book, perhaps you will change your eating habits, become vegetarian or vegan. Or not. You may even become an advocate for reforming how we treat wildlife, farmed animals, animals used in research and animals used for sport. Whatever decision you make, as long as you feel better informed than you were at the outset then the book has been a success.

A note on the text

There are numerous references in the text to the various nations of the United Kingdom, particularly regarding legislation. Animal welfare, environmental and conservation policy are devolved matters, meaning that most recent Acts and Regulations apply only in one country of the Union. So any reference to legislation in the United Kingdom can mean listing four sets of legislation when often there are only minor differences. This is not consistent, however, and some legislation, particularly that which pre-dates devolution, may apply to England, Wales and Scotland together. Northern Ireland has had separate legislation for many years.

This is not a book about the law; to keep things simple, I generally refer to legislation as applying to 'Britain' and refer to an individual country and its legislation only where it differs markedly from the other countries of the UK.

Much environmental, conservation and animal welfare legislation derives from the European Union. Now that the UK has left the EU, there is the opportunity to make new policies and hence new legislation. At the time of writing, this process has yet to begin, and thus references in the text to EU legislation are still valid.

Finally, legislation changes frequently, whether it derives from the EU or is domestic in origin. Indeed, as I write this preface, the government has signalled its intention to make wholesale changes to current legislation that originated from the EU. One can only hope that the best efforts of animal welfare and environment campaigners prevent a substantial reduction in protection for animals wherever they are located.

Alick Simmons
November 2022

1

The Exploitation of Animals

This book is about the way that humans exploit other animals, for good or bad, and how we could and should hold ourselves better to account for that exploitation. Apart from a tiny minority, we all do it – we are all complicit in the exploitation of animals. So much so that our dependence on animals and their products has become integral to our society and our economy. That dependence has developed slowly over time, from the hunting of animals for food and skins by the first humans, to the development of pastoralism, to today's industrial-scale food animal production.

Alongside food animal production, we exploit animals for other reasons – for companionship, to further the cause of science and health, for sport and entertainment, and to conserve threatened species. Because much of this exploitation takes place some distance from where we live or goes on behind closed doors, many of us have little or no experience of or direct involvement with animals, except perhaps for a few dogs or cats. But make no mistake: it goes on, at scale, and it is increasing. More animals are exploited than ever before.

As well as tolerating large-scale animal exploitation, we tolerate different types of exploitation in some species but are aghast when the same treatment is meted out to others. There is something irrational about relishing eating a 16-week-old lamb while being appalled at the notion of consuming a six-month-old puppy.

In the developed world, we're fortunate. We have used animals as part of the fuel for the economic and social miracle that gives us the standard of living we enjoy today. For the most part we are well fed, well housed and enjoy good health. When it comes to food from animals, sophisticated rearing and efficient food processing and distribution systems mean low prices, choice and abundance. But that abundance, and our health and comfort, comes at a price. Animal use in the twenty-first century is almost inconceivable in its scale. Unless husbandry, care and our other interactions with animals are consistently good, the potential for real and sustained suffering is great.

To be clear, exploitation isn't necessarily always a bad thing. Yet although many interactions are relatively benign, there are very few that can be considered wholly for the benefit of the animals involved. Exploitation can be subdivided, according to one's values, into 'use' and 'abuse', with 'use' covering what might be considered benign interaction and 'abuse' standing for something that is detrimental in some way. It doesn't take much thought to realise that these are rather loaded terms. One person's use is another's abuse. For this reason, I believe 'exploit' is a better term – we exploit animals.

A narrow definition of exploitation covers just 'unfair or underhand use', although what types of exploitation that includes is controversial because of the way in which values vary across and between societies. Therefore, when I employ the word 'exploit', I use the widest dictionary definition, that is, we 'make full use of and derive benefit from' the interaction. It includes everything from the best to the worst, from the least interaction such as observing conserved wildlife through to more significant interventions such as eating the meat of animals and wearing their skins. And, of course, it includes interactions that many find abhorrent, such as bull fighting, and more common interactions which we choose to ignore or are generally hidden from sight. The latter includes the use of rodenticide poisons that are known to be markedly inhumane.

Some of us, including livestock farmers and racehorse trainers, make a living from exploiting animals, and some, such as riders, shooters and hunters, gain satisfaction and excitement from using animals. Some make a ritual of using animals, relying on the defence of 'tradition' to justify its continuance. Many people are comforted by the companionship of dogs, cats or other pets – a form of exploitation, albeit one that appears largely benign.

Exploitation of animals includes benefiting from medical advances that have relied on the use of research animals – for example, avoiding infectious disease by using a vaccine previously tested on animals. It includes the killing of animals for sport and the killing of other animals that are perceived to be a threat to that sport or other activity. It includes the removal of animals that are considered to be 'pests' and believed to represent a risk to safety, public health or an individual's business interests. And because we are covering all exploitation, we must also consider the conservation of animals in zoos, reserves and national parks, including where populations are monitored by trapping, ringing and other forms of tagging, and manipulated by culling and reintroductions.

For a few people, the exploitation of animals – whether for farming, hunting, research, tourism or conservation – may be their whole life. However distasteful we might find some of these activities, from their point of view their interactions with animals are simply a reason *for* being, *for* living. At the other end of the spectrum, there is a small but growing number of strict vegans who might just be able to claim that they do not exploit animals at all.[1] But for the rest of us, because of the way animals and their products have become part of our lives, it is difficult to make a similar claim.

A change of heart?

My career as a veterinarian could mark me out as being part of what might be described as the 'animal exploitation establishment'. Veterinarians working in farm practice and for the government are, arguably, part of a system that exists to make animal husbandry efficient and profitable. We manage endemic disease on farms, control and eliminate epidemic disease, and reduce the risk of exotic disease incursion and threats to public health. It would be wrong to say that we lack compassion for the animals raised for meat and milk, but that compassion is constrained within tight bounds. Sure, we care about the welfare of animals on the farm, in transit and at the slaughterhouse, all of which is governed and protected by detailed regulations. But this is designed to protect the status quo – the business of producing meat and milk – and while it would be wrong to say that government veterinarians (and, in my experience, successive agriculture ministers) were not interested in animal welfare, wholesale reform was never on the agenda.

I was a government veterinarian for over 30 years. I've had a wealth of opportunities and a wide variety of roles in that time – some of which were challenging, some of which were tough. A series of jobs like that changes you. And that's why I wrote this book. I think about animals differently now. Did I have a Damascene moment? Was there any particular event that brought about that change of heart? Was it the secondment to Australia, where I was exposed to cultural differences in the treatment of animals that almost turned me vegetarian? Or was it the 2001 foot-and-mouth disease epidemic, where it seemed the whole country was expected to support an objective of eradication to be secured by killing millions of animals? Or the BSE crisis which ran for several years from 1996, during which, truth be told, I became worn out with it all? Perhaps working with broiler chicken farmers, where the sheer scale of production was hard to grasp? Or was it leading a team collecting data to determine whether the killing of badgers was likely to be useful in controlling bovine tuberculosis?

Well, it was none of these and all of these. However, I had become sceptical of many of the interventions from early in my veterinary career. As a student, I was working in a veterinary practice in the Midlands. It was a mixed practice dealing with horses, cattle, sheep, dogs, cats and everything else. I became good friends with another student, Neil Burnie. We were passionate, enthusiastic and keen to learn, and over beer in the evening we'd discuss the day's cases and our reaction to what we had seen. Neil once said: 'I can't help thinking that most of the treatments we give are either for the convenience of the owner or because of something the owner has done.' Neil believed that much of what veterinarians were treating was a consequence of how we exploit our animals: breeding dogs for a desirable physical shape, keeping dairy cows to maximise milk yield or riding a horse over jumps to prepare for a competition. It was a profound observation for a slightly inebriated 20-year-old student. But that was Neil. In essence, Neil's argument was that, in each case, the problem is caused by how we treat the animal that we keep, whether it is respiratory problems in the inbred dog, mastitis in the dairy cow or lameness in the horse.

Inevitably, as with so many of the people I ran into during my itinerant youth, we lost contact with each other. The last I heard of Neil was from his obituary after he had drowned in Bermuda, the island where he had lived for 30 years, while working on a shark protection project. His passion for animals lives on in a foundation in his name (www.neilburniefoundation.com). And his wise observation has stayed with me.

Now I've reached a point where through experience, while I am not an abolitionist, I have become an advocate for better, much better treatment of the animals we exploit. This book draws on that experience and argues for change – change that can lead to more humane treatment of the animals we keep and interact with and, in some cases, argues for prohibition of certain practices known to be inhumane.

A sense of scale

It will be helpful at this point to consider the scale of animal exploitation. The mass of kept animals now considerably exceeds the mass of those left in the wild. For instance, farmed poultry today accounts for 70% of the mass of all birds on the planet, with just 30% made up of wild birds. The picture is even more stark for mammals – 60% of all mammals on Earth, by biomass, are livestock, mostly cattle and pigs, 36% are human, and just 4% are wild mammals.[2]

This mass of kept animals translates into huge numbers. Britain's annual production of broiler (meat) chickens is around 1 billion (1,000 million),[3] which means that on average we grow, kill and eat 20 million chickens each and every week. That doesn't include imported chicken meat – we import around 450,000 tonnes of poultry meat annually (because most imports are in the form of chicken portions, it is not possible to determine the numbers of birds involved – but it's a lot).[4] Globally, around 66 billion birds are produced annually,[5] which equates to around 100 million tonnes.[6] The numbers of pigs, cattle and sheep reared and slaughtered in the UK are much lower, but none of these figures are trivial. We'll come back to this in Chapters 3, 4 and 5.

When it comes to 'wildlife' – or, more precisely, shooting – the numbers are similarly huge. Annually in the UK, around 47 million non-native ring-necked pheasants and 10 million non-native red-legged partridges are released into the countryside, although a much smaller proportion of these totals is eventually shot.[7] There is more about this in Chapter 6.

An unknown number of rats and mice, perhaps numbering in the millions, are killed annually using techniques many of which are known to be inhumane.[8] There is more about rodent poisons and other, mainly inhumane methods of killing rodents in Chapter 6.

In 2020 approximately 2.88 million scientific procedures involving living animals were carried out in Great Britain: 57% of procedures used mice, 14% used rats and 13% used fish.[9] I go into the detail of the use of research animals in Chapter 10.

Whether it is farming, research or wildlife, the sheer scale of these figures should give us all pause for thought, for two reasons. First, you risk slipping into thinking that these are just numbers and not individuals; if we assume that each of these animals has the capacity to suffer, rounding the figures into millions doesn't change that. Second, even if, say, only 0.5% of these animals suffers in one way or another, that's a lot of suffering.

A moral imperative

The risk of individual suffering and the potential for amplification of that suffering as a result of the scale imposes on us a moral duty: do we, as the beneficiaries of animal exploitation, have an obligation to make decisions, that is, to get this right? Or do we not bother and stop caring, carry on and leave everything to trust? Do we walk away and have nothing do with animals and their products in any circumstances, or do we try to make and influence better decisions?

If we believe in better decisions, in essence there are two questions here: we need to ask ourselves 'whether?' but we should also be asking 'how?' That is, as well as doing our best to decide whether any use of animals is acceptable, we should also consider what types of animal exploitation, what procedures and what privations are acceptable. As we will see, although this involves evidence-based decisions that you as an individual are encouraged to make, there is an argument for bringing in wider society to influence the outcome.

Answering these questions is a matter of ethics. Can the inevitable impact of human intervention on each and every animal ever be justified? And if the answer is 'yes', under what circumstances? Of course, circumstances which might be acceptable to you might be unacceptable to me, and in some, perhaps many, circumstances the decision might best be left to the individual. On the other hand, the nature and scale of the exploitation might lead society to reach a conclusion that legislation to prohibit or license that activity is necessary, or that other sanctions are justified.

Animal exploitation can be defended in two main ways:

- First, because wild animals can suffer terribly – starvation, predation and exposure to the elements are major causes of mortality – why bother to protect the animals we exploit? The answer is simply this: humans are moral beings, and morality brings with it responsibilities, and this extends to the animals with which we interact. Because an impala is pulled limb from limb by a pack of wild dogs, that does not mean we can abrogate our responsibilities to the animals with which we interact.

- Second, by breeding them for food or research, these animals are given the 'gift' of life. Isn't that enough? Again, given that the animals have not been given the choice and there is evidence of their capacity to suffer, as soon as we take control, we take on a duty of care.

This brings us to consideration of 'abuse'. I'm clear where abuse of animals starts, or so I used to think. Yet as soon as you gather a little knowledge and apply a little thought, it is less clear-cut than we might suppose. The line between 'use' and 'abuse' is not easily drawn. Is it acceptable to grow broiler chickens in groups of 20,000 and more, housed at a density of 20 birds per square metre, when in the wild they might live in groups of 10–15 spread over hundreds of hectares? The farmed birds might grow quickly, mortality might be very low, and the enterprise might be profitable, but at what price to the birds? On the other hand, perhaps the chicken is so lacking in intelligence that inhibiting its natural behaviour through environmental management and genetic selection means little or nothing to the individual.

Whether we accept that animals, particularly those kept under close confinement, have behavioural 'needs' is a key issue in animal husbandry. One can imagine that if those needs are genuine then keeping animals in a barren and unstimulating environment is potentially a source of considerable suffering. The behavioural needs of animals will be discussed in more detail in later chapters.

Is it acceptable to perform invasive neurosurgery on a rhesus macaque to insert electrodes in its brain and perform experiments on it, intermittently, for several years, if the results of the research lead to more effective treatments for dementia? What if the research is basic science and simply intended to provide a better understanding of how the brain works, with no immediate practical application? Would you prohibit the latter but not the former?

Wherever the intervention lies on the spectrum between good and bad, there ought to be an abiding concern about the welfare of animals in almost every circumstance where humans intervene. As we saw earlier, animal exploitation is not confined to agriculture and research. It includes anything where humans intervene in the life and fate of animals. Animal exploitation, therefore, includes the keeping of animals in zoos, as pets and companions, intervening against 'pests', conservation activities, the impact of the built environment, hunting and trapping, wildlife 'management', and any other activity where an animal's life and experience are altered – and not in every case for the worse.

In choosing to rely on animals for food, clothing, companionship, sport or anything else, we are accepting and condoning some degree of intervention in their lives. This might be direct intervention – I adopt a stray cat, you shoot a pheasant, he keeps cattle. Or perhaps it is indirect – I buy leather shoes, you go to the races, she gets a new heart valve.

Each of those interventions has an impact. The animal might be killed prematurely so we can eat its meat, it might be prevented from breeding and displaying its full range of normal behaviour, it might be mutilated (for example, castrated), it might be trapped, poisoned or shot, or it might be surgically implanted with a device to aid scientific research.

Whatever the nature of the intervention, I believe concerned people ought to give it some thought. Like it or not, we each have a stake and hence an interest in farms, zoos, households, the countryside, research facilities and other places which keep and exploit animals even if we are not directly involved in their care, management or survival. And however you care to describe it, these animals are exploited; the relationship is inevitably one-sided, with few of the benefits falling to the animals. We can and should look after them well, use humane means when killing is unavoidable, and rear and eat fewer of them. The only alternative is to stop exploiting animals altogether – but how realistic is that?

The development of values

Assuming we choose not to stop making use of animals altogether, we have a moral responsibility to make rational decisions about animal exploitation. That responsibility obliges us to be better informed. Better information isn't always easy to come by, but that shouldn't be an excuse for making arbitrary decisions.

Periodically I run into people who have made decisions about food and animals and what they're prepared to accept. It's not uncommon for these decisions to flow from almost no information, or to be based primarily on some ill-informed posting on social media. We can do better than this. It might mean a little bit more effort, but it's worth it.

I advocate developing your own values based on hard information. It pays to think about it and keep thinking about it. It's the way I did it – although I did have access to good information. I forged a career from animal exploitation, so there has been ample opportunity. Whether it was surgically neutering a cat to prevent it breeding or controlling avian influenza (AI or bird flu) in enormous flocks of poultry, it is clear that much of what I did was to facilitate more efficient exploitation of animals. Arguably, a cat does not directly benefit from being spayed, and the chickens might well have avoided bird flu had they not been kept in such large groups.

I am not recanting. To be clear, as long as we intervene in the lives of animals, then a substantial proportion of those interventions will be for our convenience. But being uncomfortable with one type of animal exploitation should not automatically mean that the rest is beyond the pale.

I am not advocating veganism or adopting an absolutist position but, rather than shrugging it off, I am an advocate for intelligent and well-informed decisions. In some cases I will accept certain types of exploitation. In others I won't. Accordingly, I still eat meat, albeit less and less, and certain types I avoid altogether. I go fishing (admittedly with not much success), I've killed animals for food and to relieve suffering, and I have directed mass killing of farm animals to control epidemic disease. I've ridden horses and I keep cats. My very survival, following a life-threatening illness, was ensured by the results of experiments on animals, many of which were completed decades ago. But I am a passionate conservationist, birdwatcher and wildlife photographer, and I have a strong interest in animal welfare and animal ethics. Having recently retired from public service, I am in a better position to do something to further these interests. This, in part, explains why I have written this book.

Now that I am no longer working full time, I agonise over this issue constantly. Why are there inconsistencies in my behaviour? Why worry about badger killing and its humaneness when I go fishing with barbed hooks? Should I become vegetarian and campaign against animal experimentation? I sometimes mull over the things I've done, seen, colluded in and failed to stop. Despite that, over the years and almost without conscious effort, I've developed values that set the boundaries of the nature, extent and duration of animal exploitation that I am prepared to accept. These values, shaped by experience and knowledge, are not fixed. Influenced by events, new information and science, and in discussion with experts and friends, I keep them under review and, occasionally, they change. You could call it a personal ethical framework – and I provide more detail about this in Chapter 11.

If you are reading this book, it is likely that you similarly are interested in animals and their welfare. You might have questioned whether it is right to eat meat, and you might worry about whether animal experimentation is justified. You might have qualms about certain types of livestock farming and be sceptical about fox hunting or fishing. There is every chance you have developed a set of values, a personal code that establishes your own boundaries for animal exploitation, even if you don't call it that. But, like the majority of people not directly involved in animal exploitation, getting information to set those boundaries is hard. How can you be sure your code is informed by fact and not overly influenced by others' dogma and propaganda? Is it consistent, defensible and coherent? Is it up to date? Are you confident that you have complete and accurate information? Or do you worry that have you been captured or hoodwinked by pressure groups and commercial interests?

Thanks to ever-increasing knowledge and transparency in our society, we are better able to scrutinise and regulate animal use than ever before, should we choose to. As a result, you would expect that the citizen was able to make granular choices based on information from a variety of sources. And this is in addition to any regulatory framework that governs the use of animals. However, the ability to scrutinise is neither consistent nor uniform. Much animal exploitation takes place behind closed doors or, on the grounds of security, in secret, and is simply unaccountable.

Some people choose not to look; perhaps they worry that if they visited an intensive poultry farm or a slaughterhouse, they might not like what they saw. However, for those of us committed to informed choice, despite living in a democracy, exercising that choice is all too often hindered by a lack of transparency and public accountability.

Where do you start? Let's start at the extremes, at the very boundaries of a spectrum of ethics. At one end is the absolutist position which has it that any and all animal exploitation is wrong since it will inevitably cause the animal to suffer at some stage. It holds that in an environment which we share with animals, most and perhaps all interactions and interventions between animals and people should be prohibited or avoided. It is a rational position, but it is obviously inflexible. Of course, this includes veganism and an avoidance of all animal products including leather, down jackets and the new banknotes finished with animal protein as well as meat and dairy products. I can appreciate the motive but it strikes me as too rigid a position as it takes no account of the enormous differences between the best and worst of animal care. Further, it ignores the benefits. For instance, the absolutist would have us keeping no pets. I am really not comfortable with that. I gain a great deal of pleasure from my cats and, subjectively, I think they gain pleasure from me.

The absolutist has no truck with animals in research whatever the objective. This is difficult territory. Few of us, I believe, would defend the use of animals to research the safety of cosmetics. Our lives do not depend on the development of a new skin moisturiser, particularly if ensuring it is 'safe' involves using several hundred rabbits. And in any case, the alternatives to using animals to determine the safety of new cosmetics are effective in ensuring safety.[10]

What about research into disease? The absolutist, abolitionist view dismisses the benefits that accrue – for example, the human (or animal) lives saved by dint of increased scientific knowledge. Many of the significant advances in medicine, including vaccine technology, cancer treatment, organ transplantation and cardiac medicine, would not have been possible without using animals at some stage. Although better nutrition and hygiene have contributed, the sustained reduction in child mortality and our increasing longevity are thanks in no small part to the results of research on animals.

Increasingly there are alternatives such as cell culture which allow for research using sophisticated systems that recreate conditions necessary for research but in 'glass dishes' (*in vitro*) rather than in a live animal. However, the alternatives have limitations; complex problems require complex solutions, so researching, for instance, heart disease – where in most cases the entire circulation has to be taken into account – inevitably involves live animals.

At the other end of the spectrum there is the argument that animals have no value beyond being a commodity – as a piece of meat, as a research tool, as a hunter's quarry. Suffering does not need to be considered, since animals cannot suffer – or the suffering endured is overridden by the perpetrators' pleasure, wants or needs. Thankfully, few people now subscribe to such a position.

Between these two extremes lies the utilitarian argument. Utilitarianism is the doctrine that actions are right if they are useful or for the benefit of a majority, and it would hold that some but not all animal exploitation is acceptable. Hence, animal experimentation is justified provided the research advances, for example, medical knowledge and no alternative exists. It follows that the minimum number of animals must be used and that the procedures must be carefully regulated to minimise suffering. A more difficult question is whether it is justifiable to use animals in the quest for knowledge and with no immediate prospects of a practical application – known as basic research. I will return to this in more detail in Chapter 10.

Another difficult question concerns food animal production. Can a utilitarian argument be made for eating meat? There are alternatives to animal protein, so why rear animals, often in conditions detrimental to their welfare, so they can be killed and eaten? I explore the farming of animals in Chapters 3, 4 and 5.

There are no easy answers. We need the ability and knowledge to make a choice from a wider appreciation of how animals are kept, cared for and killed. And from an appreciation of what we as individuals and members of society gain as a result. In other words, there is a balance to be struck. Striking that balance requires each of us to be better informed.

Should we treat all animals the same way?

For reasons that I hope are obvious, I don't subscribe to an argument that animals can't suffer or, importantly, that only certain species have the capacity to suffer. There is an enormous variety of mammals, birds, reptiles, fish and invertebrates. They exhibit a wide variation in anatomy and physiology, particularly of the nervous system. Physical differences can be huge and the way that different species behave varies considerably. There is, I believe, a natural tendency to believe that animals that share a broadly similar anatomy with ourselves are more intelligent, and have a greater capacity to feel and to suffer, than those that are very different. Hence we assume that chimpanzees are capable of feeling pain and suffering. Most people would assign the same capacity to most other mammals, although to varying degrees; and when it comes to mice and rats it is not consistently applied. We as a society tolerate or, perhaps more accurately, choose to ignore the most extraordinary but generally legal abuse of intelligent rodents when control of their numbers is deemed necessary. We use manifestly inhumane poisons and inhumane traps to kill the mice in our kitchens. And yet any suggestion that similar types of traps, albeit scaled up and more powerful, could be used for killing surplus populations of deer that are threatening forestry would have many of us up in arms.

The evidence that chimpanzees have complex societies, can make and use tools and show intuitive learning is somehow not surprising. Much of the knowledge about chimp learning and tool use derives from studies in which the chimps are rewarded for problem solving. In his book *Are We Smart Enough to Know How Smart Animals Are?* primatologist Frans de Waal describes how it was concluded that various species of gibbon were 'backward primates' because they failed to solve problems that chimps could easily complete.[11] This view didn't change until a different approach was taken, one that took into account that gibbons were exclusively arboreal and hence were anatomically quite different to chimps. By a simple redesign of the problem, which involved placing the object central to the study in a hanging position rather than lying on a horizontal surface, it was established that the gibbons were capable of solving many of the problems previously solved by chimps.

De Waal argues that we need to think smarter if we wish to understand the extent of animal cognition, and we are destined to fail in that endeavour if we make assumptions about the behaviour of animals. There's a lesson here for those with an interest in animal behaviour: making assumptions about an animal's needs solely from extrapolating what we know about ourselves and from other better-studied animals is likely to lead to serious mistakes when it comes to determining their needs and wants. And as science pushes back the boundaries of our knowledge and we understand better the behaviour of the less well-studied animals, it is becoming increasingly clear that some animals with a very different anatomy, behaviour and physiology to those with which we are more familiar, such as lobsters and octopuses, are very far from being automata.

We should not therefore treat all animals the same way, but striving to gain a better understanding of their behaviour and needs will ensure that we are better placed to meet the challenge of providing them with a safe and stimulating environment. This brings us to a consideration of whether animals can think, feel and suffer.

The capacity to suffer

A great deal of effort has been expended, by scientists and philosophers, in an attempt to determine whether animals have the capacity to suffer. As yet there is no universally accepted definition of pain, although this one from the International Association for the Study of Pain is widely used:

> an unpleasant sensory and emotional experience associated with actual or potential tissue damage, or described in terms of such damage.[12]

However elegant this definition, it does not solve the problem. It's effectively impossible to know whether animals feel pain and have the capacity to suffer, because pain (and pleasure) are subjective. You cannot be certain an animal is suffering, frightened or in pain. It is equally difficult to determine whether an animal is experiencing pleasure. This is because pain and other feelings are experienced only by the subject, and it applies even to other members of our own species. If I fall and break my leg in front of you, I might howl and writhe and you might conclude that I am in pain. But you don't feel my pain. You have simply extrapolated from your experience of pain, seen how I have behaved and concluded, 'He is in pain'. Making that assumption is the basis of empathy and compassion in society.

It's no different with animals. A dog with a broken leg might limp and howl. Although we can never be certain, for the same reason that you would, I hope, empathise if the same accident befell me, most people would conclude the dog was in pain and want to relieve its suffering. It wasn't always that way. The seventeenth-century French philosopher René Descartes denied that animals had reason or intelligence; in effect animals were automata. He argued that animals did not lack sensations or perceptions, but these could be explained mechanistically. In contrast to humans, animals by virtue of not having a soul could not feel pain or anxiety. If animals showed signs of distress then this was to protect the body from damage, but the innate state needed for them to suffer was absent.

Descartes' views became widely accepted in Europe and North America, allowing animals to be badly treated with impunity, and it was not until the middle of the nineteenth century that society's view and ultimately the law changed to provide a modicum of protection. Those laws have changed over time, and now in Britain and the rest of Europe there are comprehensive laws that govern our use of animals. Current legislation in Britain provides general protection through a duty of care, while more specific and detailed legislation covers farm, research and other animals. However, as we will see, these laws are inconsistent and in need of amendment to take account of our increasing knowledge of sentience and the capacity of animals to suffer.

The eighteenth-century philosopher Jeremy Bentham was influential in changing the way we treat animals. He argued that the question is not 'Can they reason?' nor 'Can they talk?' but 'Can they suffer?' In other words, in the absence of better information, animals need the benefit of the doubt. And despite an enormous continuing research effort we are still no closer to a conclusion about consciousness in animals.

In 2017, the animal welfare scientist Marian Stamp Dawkins argued that 'although the pursuit of consciousness … is one of the most fascinating in the whole of biology, the extreme difficulty of the search means that understanding is still a long way off'.[13] She contended that we should not 'base the science of animal welfare on the assumption that we understand consciousness or can decide which species are or are not conscious. Animal welfare is far too important to be made to wait until the hard problem of consciousness has been solved.' She suggested 'two criteria – what keeps animals healthy and what they themselves want – that together constitute a necessary and at least partly sufficient basis for an objective, consciousness-free science of animal welfare'. In essence, Dawkins presents a more modern argument for giving the benefit of the doubt to animals and their welfare.

But to which animals should we give the benefit of the doubt? Which animals have the capacity to suffer and which do not? Surely there must be some animals that we needn't worry about, species that are so poorly developed, with such a small brain and so lacking in intelligence that they lack any capacity to suffer. For other animals, the smart ones with a large brain, if society deems it necessary or desirable to exploit them, we will look after their needs and strive to stop them suffering. The rest of them, those creatures with poorly developed central nervous systems and that don't look at all like us, we can stop worrying about and do the minimum necessary to keep them alive. We can exploit them to a conclusion – a fish pie or a scientific paper, for instance. Surely science can give us the answers we need? 'We will allow these animals to suffer, but not those.' If only it were that simple.

To play safe, we could apply to all animals Dawkins's plea to give animals the benefit of the doubt. That way, you can avoid any risk of suffering by avoiding all interactions with animals. An alternative, slightly more flexible approach would be to avoid eating 'anything with a face'. That means you'd still be able to eat shellfish, but you'd have to avoid fin fish and crustaceans as well as mammals and birds. Is the presence of both a nervous system and an anatomy similar to humans sufficient to conclude that an animal is capable of feeling pain (and, in this comparison, the anatomy of a fish sufficiently similar to that of a human – in comparison to a cuttlefish or a cockle)?

In an important paper published in 2014, Sneddon and colleagues reviewed the evidence of pain in animals and concluded that although it cannot be proven that animals experience pain, it also cannot be proven that they do not.[14] Fish, cephalopods and decapods all demonstrated responses in experimental conditions that were analogous to responses that are well recognised in mammals and birds and which are generally accepted as evidence of a pain response. This gives the lie to the argument that only vertebrates are capable of feeling pain. In contrast, arthropods (mainly insects) do not exhibit similar responses.

In Britain, the use of cephalopods was, in 2012, brought within the scope of the law governing animal research – although, curiously, and despite the evidence, neither decapods nor cephalopods are included within the scope of the general animal protection law.[15] However, the recently enacted Animal Welfare (Sentience) Act 2022 includes decapods and cephalopods, paving the way for regulations that would protect these animals in a similar way to vertebrates.[16]

Sentience

Animal sentience refers to the ability of animals to feel and experience emotions such as joy, pleasure, pain and fear. The degree to which animals feel these positive and negative states is contentious. While there is consensus that most mammals are sentient, or at least they are given the benefit of the doubt, the extent to which the same view is applied to other groups of animals such as birds, fish, and reptiles remains the subject of vigorous debate.

The belief that animals have the capacity to suffer drives the animal welfare movement. It is the reason why animal protection laws exist. But, in addition to the capacity to suffer, should we assume that these same animals have other feelings? That is, the ability to feel, perceive or experience? In other words, are they sentient?

There's a difference between sentience and the ability to think and reason. It's clear that vertebrate animals, or most of them, can think. I've watched my cat stalk a vole in the overgrown field behind my house. It is clear, at least to me, that a stalking cat is making decisions continuously. The vole is buried deep in long grass and is stock still. It is making no sound but the cat can smell it.

Should the cat continue, or is the effort not worth the prize? Should the cat dive into the rank grass following the scent trail of the vole, or simply wait to see if it reappears? That looks to me like the cat is thinking, all the while weighing up the risks, effort and potential reward. Others might argue that the cat is simply following a set of inbuilt instructions, that is, acting purely instinctively. That argument falls over when you consider how a kitten's behaviour changes as it gains experience, learns and grows. At the age of 16 weeks or so, a kitten will chase and pounce on anything that moves, but with experience it will become much more selective and ignore those stimuli that don't lead to something worth hunting. It now takes a great deal of persuasion to get my middle-aged cats interested in anything moving unless it leads to an early meal. Clearly, cats can learn.

But what, if anything, do cats feel? Let's return to my hunting cat – and assume that it is experienced and wants to catch something tasty. Successful or not, does the cat feel exhilarated, frustrated, pleasure or nothing at all? In over 30 years of keeping and observing cats, I have concluded that cats are not particularly smart – they are extremely finely tuned hunters but seem to exhibit limited capacity to learn from experience in the same way that, say, a raven does. But that is not to say that we should assume that the cat, or any other mammal, does not have the capacity to feel.

These feelings, if indeed that is what the cat is experiencing, can be described as sentience. If we assume the cat is sentient, then it follows, in the absence of other evidence, that we should assume the other animals we are responsible for are sentient too. And if these animals are capable of feeling pleasure and frustration, then we have a duty to consider whether these feelings have a bearing on animal welfare.

Animal rights

I've touched on the responsibilities of animal keepers and anyone else with influence over the way in which humans and animals interact. These responsibilities, set down in Codes of Practice, Acts of Parliament or in your own moral code, are things you must or must not do. You must provide the animal with food and water, protect it from disease and injury, keep it comfortable, allow it space and a suitable environment to behave naturally. You must kill it humanely. You must not beat it, neglect it or confine it in too small a space. But meeting these responsibilities alone might not be enough. Do animals have rights? And if so, which ones? Do animals have a right to life, for instance? Remember that while the evidence is equivocal, we take the view that most animals have the capacity to suffer.

Peter Singer's book *Animal Liberation* had a profound effect on me.[17] I read it in 1989 while on secondment to the Australian government, at a time when I was already having doubts about our relationship with animals.

Although the book has been revised several times since its first publication in 1975, Singer's basic premise remains constant: treating animals differently from humans is 'speciesism', and we 'should give up our claim to "dominion" over the other species of animal' and cease keeping animals for meat, research and other forms of exploitation. It is a compelling argument and one that has become more relevant as our knowledge of self-awareness in animals has increased. However, Singer is more nuanced than the outright abolitionist. For example, more recently he has argued that a utilitarian approach to animal research may be justified provided the numbers involved are minimised.[18]

Singer also wrestles with the problem of keeping animals for food and what some describe as the 'ultimate harm', that is, killing an animal so you can eat it. While few people agree that it's acceptable to abuse an animal while it is reared for meat, some people suggest that it is hypocrisy to argue for better conditions for farmed animals when the ultimate aim is to kill and eat it. There is also an argument that if it were not for farming and the demand for meat, many of these animals would not exist – so eating meat is good for animals. There is no right answer to this – where you land is a matter of personal ethics. However, even if you were to argue that the few weeks or years of life for a chicken or a bullock is justified before it is killed so we can eat it, it surely means that the conditions for that chicken or bullock must meet their wants and needs – and, as we will see later, many systems of farming fail to do so.

My personal position has changed over time. While I avoid certain meats (for environmental as well as animal welfare reasons) and I remain opposed to the use of non-human primates in science, I still eat some meat and support the use of some animals in applied medical research. It's a utilitarian position. And yet I have been described as an animal rights activist (or even extremist) because I balk at the routine killing of wildlife using techniques that are demonstrably inhumane and for which there is no evidence of benefit.

My preference is for decisions that affect the animals we exploit to be backed by evidence and basic principles. In some cases, I will conclude that the exploitation is acceptable with safeguards, and in others that it is unacceptable regardless of any of the safeguards that might be applied. Is that the position of an advocate of animal rights? Or of someone who believes we have responsibilities towards animals? Or a bit of both? I'm not sure. I don't believe that a firm distinction can usefully be made between animal rights and our responsibilities to animals. Which means that ascribing labels to particular positions is probably not helpful.

Is there a line to be drawn, and where do we draw it?

Do these arguments, including Singer's, form a case for veganism? That is, should we forgo all exploitation of animals, including any direct or indirect involvement? Or, though less likely if you are reading this book, you may follow Descartes and deny the feelings of animals. It depends on your values. A more rational position, I believe, lies somewhere between these extremes. That is, we apply a code of ethics based on an acceptance that many, perhaps most animals are sentient and have the capacity to suffer, but that some exploitation is acceptable on utilitarian grounds. The spectrum is very wide and leaves a great deal of room to compromise (or wriggle room, depending on your viewpoint). I don't seek to dictate where you sit on the spectrum – I would rather it is not at either extreme but that's your choice. Where you sit has to be a personal decision. There are good reasons for this. First, you should think for yourself. You might conclude that it is unseemly for morals to be dictated by anyone but yourself. Second, it's complicated. There are so many ways in which animals are exploited, and some will be of more importance and immediate to you than others. Third, if you do it properly, you'll keep it under review.

If you are to make informed choices, then you need information. There is no shortage of that. You can lose yourself on the internet for days at a time going from one extreme to another. At one end of the spectrum, there's any number of detailed and sometimes lurid websites run by earnest and often well-informed animal welfare organisations seeking your endorsement, money or a change in your behaviour. At the other, there are farming and food organisations and businesses extolling the welfare virtues of their farms, the care lavished on their animals, and the quality of the meat, eggs and clothing that they produce. You could read the animal welfare regulations covering farm animals, pets, research animals and wildlife. You can gain access to a wide variety of scientific literature. There are organisations like the 'Red Tractor' scheme and the Royal Society for the Prevention of Cruelty to Animal's 'RSPCA Assured' scheme, both of which approve farms that either meet statutory requirements or are run to a higher standard that is claimed to ensure better welfare.[19]

All of these groups and representative bodies are seeking your attention. On the one hand, there is propaganda trying to change your mind about eating meat and urging you to become a vegan, or encouraging you to campaign against fox hunting or to eschew that new pair of leather brogues you had your eye on. On the other, we are told 'Trust us, this meat comes from farms run well by people who care.' Even if you give all of them the benefit of the doubt, how do you find a way through it, distil it down and make an informed choice? It's difficult, and if you are just a little sceptical about the claims, it becomes close to impossible.

Consumers making decisions about animal exploitation is only part of the solution. Let's not delude ourselves: my decisions about what not to eat, wear or shoot will not make a great deal of immediate difference. I made a decision several years ago not to eat pheasants and other 'game' birds because of my concern about the way in which organised shoots arrange for the arbitrary, routine and inhumane killing of predatory mammals and birds. But I am not naive. That decision isn't going to make much of a difference to the fate of the stoat or the carrion crow that fetches up on a shoot looking for pheasant poults. It might not even make much difference if lots of people make the same decision.

But significant change can come eventually. Word of mouth, the efforts of the conservation and animal welfare bodies I support and the exercise of consumer choice all have an effect. Changes to animal welfare law in the last 60 years have been profound. This did not come about because parliament simply thought it a good idea. Research and public inquiries don't begin spontaneously. Public pressure for animal welfare reform made parliamentarians take note – public debate, scientific inquiry and legislation followed.

Change does not come solely via legislation. Processors and retailers, especially the big supermarkets, study consumer behaviour carefully and change their offer accordingly. Perhaps the apparent increase in the number of bags of pheasant carcasses found dumped in field margins is indicative of a change in purchasing habits; perhaps supply is outgrowing demand. Perhaps the increasing range of vegan and vegetarian food in supermarkets is further evidence of the collective impact of individual decisions.

Although personal decisions made by consumers may be effective if sufficient numbers make the same decision, there are limits. You cannot 'vote with your feet' when it comes to the use of traps to kill wildlife or the use of animals in research. Despite the general public's overwhelming opposition to the use of snares to catch wild mammals, they remain largely unregulated, freely available and widely used.[20] This continued use of snares is a consequence of a combination of three factors: first, the lack of an evidence base about their animal welfare impact; second, the influence of those who seek to maintain their use; and, third, a reluctance by governments and parliament to introduce society's values into any debate about the continued use of inhumane methods of killing wildlife.

Of course, there are advocacy groups, non-governmental conservation and animal welfare organisations all of whom can amass evidence, campaign for change and act as a collective voice for the concerned citizen. Governments, as a matter of course, seek views from relevant organisations and individuals as they develop policy and legislation, although, in my experience, members of conservation and welfare organisations are generally sceptical about the degree of influence they exert in this context.

There's a good case for some wider involvement of the citizen in decisions about animal exploitation. There is scant opportunity at present. But there is a chink in the armour. Since 2013 all premises licensed to use animals for research have been obliged to establish a body charged with overseeing animal welfare and ethics.[21] These are known as Animal Welfare and Ethics Review Bodies (AWERB). While a majority of the membership is reserved for staff of the establishment, there is a requirement for at least one lay person to be appointed. Run well, these bodies can effect real change in procedures and even stop work proceeding altogether. Their powers are limited, and inevitably the deliberations are confidential. However, despite these limitations, it is a step in the right direction in holding the actors to account better than hitherto.

Although they are protected to some degree by legislation, no such body exists for wild animals, farmed animals and indeed any other grouping of exploited animals, although some conservation bodies are making a start. Government departments frequently appoint individuals from a variety of disciplines to their expert advisory bodies, although this is generally at the national level and their involvement rarely extends to making decisions on issues of ethics. The case for independent lay persons involvement in decisions in other areas of animal exploitation needs to be considered further, and this is explored in detail in Chapter 12.

Summary

Animal exploitation forms an important part of our way of life, and contributes to our prosperity and wellbeing. The scale and nature of that exploitation means that there is considerable risk of suffering, but, for a variety of reasons, it is difficult for the private citizen to determine where, how and by how much animals are suffering. There is a moral imperative on us all to make decisions and to exert influence, but the ability to make an objective analysis is hindered because information is either scant or absent. However, better information can inform personal ethically based decisions and can be used to influence the institutions and rules that govern the way in which animals are treated.

This book is intended to help interested and concerned people to make informed decisions. It is not solely about animal welfare, and although it includes much about welfare, I have striven to avoid polemic. Nor is it a plea for veganism or vegetarianism, although it might help you make the choice one way or the other. It is not a book on animal rights, although it touches on the notion of rights for animals since I believe we need to give greater respect and protection to animals.

The remainder of this book addresses these issues, covering the various ways animals are exploited, with examples from current systems of livestock and poultry husbandry, animal research, wildlife conservation and management, and other animal uses. This includes the more insidious and indirect interventions that have a bearing on the experience and fate of animals. Chapter 2 investigates differing attitudes and anomalies in protection and care between often closely related animals. In Chapters 3, 4 and 5, I consider the main farmed species, detailing how practices both new and old affect behaviour, disease, welfare and, ultimately, mortality. Chapter 6 covers wildlife, including the dubious ethics of much wildlife 'management', and considers how best to introduce a more humane approach. In Chapter 7, I explore how the exploitation of animals in conservation is managed. Animals used in sport, whether for racing or recreational shooting, are the subject of Chapter 8. The knotty question of whether pet animals are exploited is considered in Chapter 9, and in Chapter 10 I consider animals used in research – perhaps the most highly regulated but least understood area of animal exploitation. In Chapter 11, I introduce my own personal ethical framework. Finally, Chapter 12 advocates more consistency in our relationships with animals, based on societal governance and evidence-based rules.

2

Why Aren't All Animals
Treated the Same Way?

Not all animals that we interact with and make use of have similar experiences. We exploit a variety of animals, each in a different way, for different reasons and with different objectives. Those objectives translate into different attitudes. For example, dog owners, many of whom would walk across hot coals to rescue their beloved pet, are in my experience unlikely to agonise over the use of rodent poison when mice take up residence in their homes. Similarly, the farmer who cares deeply for his cows and frets about their diet and disease, may have few qualms about killing rabbits on his farm.

I know of people who would not countenance wearing furs under any circumstances, but who insist on wearing leather shoes. I can understand why you would want to avoid furs derived from animals trapped in the wild or produced in a fur farm. But, by the same token, why buy shoes when you have no information on the source of the leather, the species of animal it was derived from, or how that animal was reared or killed? I recently bought a pair of lightweight boots to replace a leather pair that had seen better days. I was puzzled to see that the relatively small amount of leather used in the boots was described as 'yak leather'. It got me thinking. What was its provenance? Was it a wild or a domestic yak? How had it been reared and killed? Or, if it was a wild one, how had it been hunted and killed? So, I asked. I sent an email to the manufacturer's Customer Contact Centre and 10 days later I received a polite but anodyne response. It explained that the leather was derived from a domestic yak but there was no information about the method of slaughter. I suspect they didn't know. In my experience, this is a typical response. From food shops to restaurants to clothing, asking about the provenance of the animal products and how they were treated is almost always met with a blank stare. Sure, you can ask your butcher and you might get an answer. But try asking the manufacturer of your new leather shoes and see how far you get.

The different standards that prevail are fascinating. We may go to extraordinary lengths to protect a particular species from harm or extinction or to boycott certain animal products believed to be produced inhumanely, but think nothing of buying a goose down jacket despite the evidence that much of the down has been harvested from conscious geese.[1]

Why do we do that? For the most part, as we've seen, these animals, both the ones we seek to protect and the ones we continue to exploit, all have similar physiology, anatomy and nervous systems. Granted, the central nervous systems of birds and mammals are very different from each other, but there is nothing to suggest that a bird is less able to experience pain than a mammal. In the absence of better information which confirms self-awareness and consciousness in each and every species, it is safer and probably necessary to consider that all vertebrates (and some invertebrates) are capable of suffering.

Why aren't all animals treated the same way? What's the reason for these inconsistencies? Why do we eat some species and not others? Why do we tolerate the presence of some animals but not others? Why do most (but assuredly, not all) people accept the use of rats and mice in research but balk at the use of primates?

These apparently simple questions have complex answers. Even the most cursory look at our relationships with animals will uncover wide differences in the way we treat animals and differences in what we are prepared to tolerate.

There are many reasons why people either treat or value some animals differently from others. Differences in the way animals look, the physical differences between species, influence how tolerant we are of their exploitation. Similarly with the way they behave. Some differences can be explained by how familiar or unfamiliar we are with particular species, meaning that experience may make us more tolerant. However, it can be argued that the less we see, the less we care. At one extreme this engenders fear and intolerance. At the other, it leads to indifference – and this may apply particularly to animals reared in large numbers but out of our sight. Finally, our culture and history have had an important influence. Of course these categories are not mutually exclusive, but it helps our understanding to break them down and consider them individually.

For the purposes of this book, I will deal mainly with three broad groups of animals – farmed, research and wildlife – with other chapters on conservation, animals in sport and pets. This might seem arbitrary, and to run counter to the arguments I have already used, particularly when I argue that for reasons of sentience, arbitrary divisions are unhelpful. However, it is convenient, if far from ideal, to follow these groupings simply because, for a number of reasons, we as a society use them habitually.

Culture and history

Some of the differences in how we perceive and treat animals are cultural and historical. Take the horse, for example. The degree of protection afforded to the horse in British legislation and in popular culture exceeds anything that exists for other farmed animals.

The horse has always been treated differently. There are arcane and longstanding rules designed to prevent low-value horses and ponies being exported from Britain while, until recently, the government was resolutely defending the export of live cattle and sheep for slaughter.[2] There are several charities dedicated to the protection of horses and donkeys. There is no tradition in Britain of eating horse meat (at least not deliberately), while we will happily eat similar beasts such as cattle and sheep. This squeamishness about eating horses, however, doesn't seem to stop some owners from selling Dobbin to a dealer who immediately sends it to one of the two horse slaughterhouses still operating in Britain. This is illustrated by something I came across while working on meat hygiene policy in the early 1990s. There is a requirement to withdraw treatment from an animal for a specified time prior to slaughter to ensure that the residues of the active ingredient are below a legally enforceable safe level. To monitor compliance, there is a structured survey which involves the analysis of random samples taken from animals slaughtered for human consumption. On occasion, horse meat was found with levels of a particular substance above the legal threshold. However, unlike when similar non-compliances were found in sheep, cattle, pigs and poultry, there was rarely any follow-up (blatant non-compliance could result in prosecution). This was because it was believed that investigating why the horse had residues of a medicine in its carcass would be upsetting to the erstwhile owner – they would discover that their horse had been slaughtered for meat rather than being sent to the promised 'home of rest for horses'.

The reason for the horse's exalted status in Britain is unclear. Perhaps it is because it was the primary draught animal prior to mechanisation, or because of the role it played in imperial wars, or because of its status as a plaything of the rich and powerful. Perhaps it's all three and more. It doesn't really matter why. Even if the reasons are lost in the mists of time, it is a fact: most people care more about the welfare of horses than that of cattle and sheep.

Differing attitudes to wild animals and their perceived suffering at the hands of humans also lead to differences in tolerance. The controversy that rages around fox hunting is due in large part to differences in how the suffering of the fox is perceived. The internet is full of references to the proponents of fox hunting asserting that the fox is killed instantly by the hounds by way of a 'quick nip to the back of the neck'. By virtue of my career, I have been involved in the killing of many different species of animal using a variety of methods. I am quite confident that at no point would I have considered that a 'quick nip to the back of the neck' was an effective and humane method of killing any animal. But then I have a different cultural background and experience to those who hunt foxes. In a report for the League Against Cruel Sports (LACS), the forensic pathologist Professor Ranald Munro backs up my view by stating:

> Post mortem examinations of hunted foxes highlight the barbarity of fox hunting. The proponents of fox hunting claim that the fox dies quickly following a well-judged bite, by the lead hound, and a shake of the neck. This is a myth. The reality is that the fox is seized by the hind legs, or over the back, and then dies as he is ripped to pieces by the hounds.[3]

Similarly, farmers clearly care about the welfare of their stock, but – perhaps because of their day-to-day experience – their perception of the pain and suffering that animals should tolerate is generally different to that of the rest of the population.

Between cultures, the differences appear greater still. Two years after joining the UK government's veterinary services in 1985, I was seconded to the Australian government's veterinary services. One experience remains fresh in my mind. I was assigned to a slaughterhouse in New South Wales which was processing some 3,000 sheep every day, and one of my duties was supervising the unloading of the sheep from the animal transport vehicles. Given that these were generally elderly ewes it was not uncommon for one or two to arrive dead (which in itself was cause for alarm) or to be recumbent. I would expect that a recumbent ewe that was unlikely to recover (they rarely did) would be quickly dispatched and not enter the slaughterhouse since it was clearly unfit to be eaten. Instead, I found a number of recumbent and clearly dying ewes next to where an operative was skinning an already dead sheep. I asked why he had not used the humane killer on the dying ewes. His response was 'If I do that now, by the time I get to them they will be stiff and too difficult to skin.' It didn't take long for me to make him change his mind.

Now this was an isolated and extreme incident and not typical of my experience in Australia. But it would be fair to say that attitudes to farmed animals in Australia tend to be harder than those in Britain. For instance, there were numerous routine procedures carried out on cattle and sheep in Australia, from dehorning to castration to mulesing (see glossary on page 238), that in Britain were either permitted only under anaesthesia or prohibited. At the time, the value of a sheep or a cow was approximately half what it was in Britain, and animal products were cheap. I mused about whether this difference in value explained the more casual attitude to animal welfare, but I am inclined to believe that much is simply due to a divergence in social values which has led to the development of very different attitudes to farm animals.

A few years later, I was birdwatching in Extremadura, Spain. We were on the trail of the great bustard when we stopped for the night in a rural bodega. While we had a meal and a beer in the restaurant the owner switched on the TV. It was showing a bullfight. As I ate my steak, I probably saw the ritual in greater detail than if I had attended the bullring in person. Telephoto photography gave us all close-up views. It was pretty unpleasant to watch but I was able to see the process of successive injuries that are inflicted to present the matador with the opportunity to sever the bull's spinal cord. What was more striking was the reaction of the other customers. It varied between mild interest and apparent indifference. They might have been watching *Strictly Come Dancing*. What to most of us Brits was inexcusable cruelty was routine, part of daily life, part of the TV schedules, and not much to get excited about.

The dog, an animal which in many British households has achieved a status similar to that enjoyed by the cat in Ancient Egypt, is considered in most of Asia to be a dirty, disease-ridden animal. An Indian national would be horrified if you mentioned that your dog enters the house and sleeps on your sofa. He or she would be speechless if you admitted that it sleeps on your bed. I don't have a dog and, if I did have one, it wouldn't be allowed to sleep on the bed. On the other hand, if I am in bed feeling a bit under the weather it is surprising how easily my cats find a way in.

At the other extreme, in some cultures, dogs are essentially farmed for meat. While the idea of eating a dog makes me queasy, it is my cultural heritage that evokes that reaction and not a science-based rationale. Objectively, if I am prepared to eat beef, and if both the dog and the steer are reared in conditions that meet the animal's needs and both are killed humanely, it is hard to argue that eating a dog is wrong, however much you find the idea abhorrent. There is a similar argument which involves a larger animal, the minke whale, as we move from domestic animals to consider those that live in the wild.

Britain once had one of the largest whaling fleets in the world and played a substantial part in driving most of the species of great whale close to extinction. In 1982, the International Whaling Commission (IWC), the supranational body that regulates the whaling industry, agreed to a moratorium on commercial whaling. That moratorium stands to this day, and although some countries have withdrawn from the IWC or have exploited loopholes in the agreement to allow continued whaling, the populations of many threatened species have begun to recover. Smaller species like the minke whale which mature more quickly and are relatively fast breeders have recovered more quickly than the larger, slower-maturing species such as the blue whale.

A rat is a rat is a rat

Some of the differences in how we treat animals arise from the circumstances of their exploitation. That is to say, different attitudes seem to depend entirely on how the animal is exploited. The brown rat (*Rattus norvegicus*) is one of the world's most widespread and successful animals. Thought to have originated in northern China, it is now established on every continent except Antarctica and is frequently a commensal of humans. That is, like the house sparrow, its fortunes are often closely linked with human activity and the built environment. It has been associated with the spread of numerous diseases and is generally considered a 'pest'.

We come across the brown rat in three main circumstances. It is often kept as a pet, it is used widely as a laboratory animal, and it is common in the wild, where it is generally described as 'vermin' and treated accordingly. There's no reason to believe that the rat's ability to suffer is any different in these markedly different circumstances. Yet the pet rat is cosseted, comforted and petted (and is frequently obese), the laboratory rat is kept in a highly sophisticated and regulated environment, and the wild rat runs the risk of being deliberately, slowly and painfully poisoned or being inefficiently and incompetently trapped using unregulated and unsupervised equipment.

It is easy to conceive of a household where all three interactions occur. A teenage boy keeps a pet rat and will happily spend substantial amounts of his parents' money to treat its respiratory problems. His older sister is a scientist in a laboratory where disease is deliberately induced in rats in a quest to better understand similar disease in humans. Their father regularly puts down rodenticides because rats are raiding the feed stored for his racing pigeons.

The differences in the way in which we keep or interact with the brown rat are reflected in our legal frameworks. The pet rat is protected by general animal welfare legislation, the Animal Welfare Act 2006 in England and Wales (there is similar legislation in Northern Ireland and Scotland), which imposes a duty of care on the person in charge of the animal. There is comprehensive and prescriptive legislation that governs the life and experience of the laboratory rat. But there is minimal legislation to protect the wild rat despite the many methods we have devised for killing them – few, if any, of which can be described as humane. Ultimately, these differences reflect our different attitudes to animals in different circumstances. Yet, when examined dispassionately and taking into account that rats, wherever they are, have the same capacity to suffer, we surely have to question why this is the norm.

The UK government's position is that it 'opposes all whaling apart from limited aboriginal subsistence whaling', and that 'whaling is unacceptably cruel and is economically unnecessary.[4] That is a clear and unequivocal statement. But is it consistent with the other government positions on animal welfare? The population of minke whales is estimated at 500,000 and it is classed by the International Union for the Conservation of Nature (IUCN) as being of Least Concern.[5] A stable or increasing population of a species of whale, like those of most deer species in Britain, might be suitable for exploitation. Of course, one would want to ensure that, as with any other wild animal, minke whales were killed humanely. The evidence is that there is no consistently used humane method of killing minke whales.[6] Therefore, the government's position seems rational. But suppose a method of killing whales humanely was widely used. If we are prepared to 'harvest' deer from healthy populations because the argument is that it can be done humanely and sustainably, there is an equally valid argument that sustainable whaling is practicable provided it is done humanely. Of course, it is very unlikely to happen even if humane killing methods were widely adopted. This is because our changed and inconsistent values are now reflected in an arguably populist government policy.

Familiarity may not breed contempt, but it drives indifference

Take the humble broiler or meat chicken. Until the 1960s it was an expensive delicacy, but chicken meat is now ubiquitous and cheap. And at the point where the vast majority of us encounter it, in the supermarket chiller, it looks less and less like a chicken. In Britain it is almost impossible to buy a freshly killed chicken, one which needs to be plucked and gutted before it can be cooked. Indeed, even the proportion of chickens sold at the next stage of processing, as 'oven-ready'– a bird which has been plucked and gutted – is falling as more and more is processed further into boneless portions or ready meals. For the majority of consumers, a plastic-wrapped package of skinned and boneless chicken breasts bears as much relation to a live chicken as the latest electric car does to the lump of bauxite which forms part of the raw materials for the car.

Let's extend the analogy a little further. You'd have a bit more under-standing of (and perhaps even empathy for) the creature you were about to eat if you'd had to rear it, kill it, pluck it and gut it before you could cook it. In the same way that your grandfather developed an understanding of his Morris Minor, because every Saturday morning he had to spend an hour or two carefully coaxing it to start, had to rebuild the carburettor every six months and spend a fortune keeping rust at bay. The twenty-first-century electric car might be more convenient and reliable but do you know what makes it go, can you fix it if it goes wrong, and, more importantly, do you care about it?

It's hardly surprising that people appear indifferent to the experience and fate of the chicken before it arrives, safely wrapped in plastic, in our supermarkets. A seemingly endless supply of cheap (and frustratingly bland) chicken meat forms an ever-growing proportion of our diet to the extent that it is predicted to become, across the globe, the most widely consumed protein. It's become a staple like wheat, rice and maize. It might even be a commodity.

Certain animal products are already fully fledged commodities. A commodity is a raw material or primary agricultural product that can be bought and sold. Examples of global food commodities are butter fat and skimmed milk powder. Both are derived from fresh cows' milk and are actively traded internationally. Beef, pork and chicken meat are slightly less developed as commodities but there are well-developed international trades in each of these meats. It follows that, for something to be traded unseen and in vast quantities, conformity and consistency is paramount. How else would you know what you were getting?

It is not for nothing that the vast majority of the day-old chicks which grow on to be the chickens we eat are supplied by a small number of giant transnational corporations. These corporations control the sophisticated broiler genetics and often directly or indirectly control the feedstuffs and the rearing environment, thereby ensuring as consistent a product as possible. The fact that much of this goes on behind closed doors adds to the sense of detachment.

That sense of detachment among the general public is so complete that even potential food scares hardly dent the sales. When the first incidents of bird flu (avian influenza, or AI) hit Britain in 2007, I was the Food Standards Agency's Veterinary Director. My team and I, having been through the bovine spongiform encephalopathy (BSE) and the 'salmonella in eggs' crises, were acutely aware of the potential effect of AI on confidence in the safety of chicken meat (and eggs). We put a great deal of effort into messages about food safety and I spent plenty of time in radio and TV studios. All the while, despite science showing that the public health risks from eating chicken meat and eggs were very low, we fretted that the tide would turn against us and there would be a major food scare. In the event, unlike with BSE and salmonella, the impact was very small and there was only a minor blip in sales. Despite AI incidents recurring every few years, sales of poultry meat continue to rise. Society has changed. Chicken meat, like bread, has become an immovable fixture in our diets. Indifference to the chicken as a living creature is almost complete.

Lack of familiarity leads to fear and intolerance

Fear and intolerance are engendered by lack of exposure, lack of knowledge and lack of experience. Some societal fears go back centuries. Despite the fact that far fewer than 1% of the British population have ever seen a wolf, and even though they became extinct here several hundred years ago, our fear endures. Danger from wolves was a common theme in early literature and folklore. Despite that, attacks by wild wolves are rare, and fatal attacks are even rarer and hard to document.[7] Contrast that with our attitude to the dog. Most people are familiar with dogs and, given the impact of dogs on our society, are surprisingly tolerant of their disruptive behaviour. It is interesting to compare that tolerance with the reaction to any proposal to reintroduce the wolf to Britain. There have been several studies into the attitudes of the public to the reintroduction of previously native carnivores (wolf, bear, lynx). In general, attitudes are generally positive but resolutely cautious. In one study, a participant is quoted as saying: 'People psychologically reject wilfully bringing something in that has the potential to bite you.'[8]

Despite widespread revulsion and outrage at the thankfully rare event of a dog killing a child,[9] as a society we seem to be quietly tolerant of the increasing number of dog attacks on generally blameless people. According to a report published in 2021, the number of hospital admissions for dog bites tripled in England from 1998 to 2018.[10] Similarly, there is concern that wolves might attack farmed animals should they be introduced, but curiously there is little popular outrage at an apparent steady increase in the numbers of attacks on livestock by uncontrolled domestic dogs.[11]

Genuine inexperience can drive fear of the unknown. The wild boar became extinct in Britain by around 1300. Since the late twentieth century escapes and releases from farms of pure wild boar and domestic pig/wild boar hybrids have resulted in a growing population at a number of sites, notably the Forest of Dean, the New Forest and along the Kent/Sussex border.

Wild boar grow large and are prolific breeders. For people not used to interacting with wildlife, they might appear to be a threat. Despite this, the numbers of people being injured by wild boar in Britain are very low (and much lower than injuries by dogs). However, this did not stop a proposal from the National Trust to kill all the wild boar at their Stourhead Estate in Wiltshire, primarily because of 'several reports of members of the public being confronted and intimidated'.[12] Analysis of 412 wild boar attacks in a number of countries suggested they were rare but that there were a number of behaviours that were likely to increase the risk of attack: for example, walking with a dog through undeveloped areas or threatening or chasing a wild pig – the sort of behaviour that people with experience of wildlife would surely avoid.[13]

Britain has a depleted fauna, and the general public has little experience of the species that have been lost – and, because of our urban lives, increasingly less experience of those that remain. However, intolerance and a desire to have wildlife either removed or behind fences is hardly likely to help us regain that experience and the ability to live, largely harmoniously, alongside the large animals that are increasingly common across much of the rest of Europe.

Attitudes can change over time

Two species of squirrel and the passions that each evoke show how attitudes to animals can change over time. The introduced eastern grey squirrel is a North American species which has become well established over much of Britain. It is cooed over by many, and fed in parks and gardens, but reviled by others because of the 'damage' it causes to growing trees, because it devours nestlings and because, unforgivably, it outcompetes the revered native red squirrel. Not only does the grey squirrel drive the red squirrel from its favoured woodlands, it spreads the squirrel pox virus. The virus causes no disease in grey squirrels but is generally fatal to reds.

The red squirrel is hanging on by the skin of its teeth in a few redoubts in England and Wales, although populations are much healthier in Scotland. There are efforts to control grey squirrel populations by shooting, poisoning, trapping and even the use of contraceptives.[14] Reports of increasing numbers of the formerly endangered pine marten, a carnivorous species that preys on grey squirrels, are cheered on by the advocates of the more acrobatic and more difficult to catch red squirrel.

I'm all for native species, particularly where they are threatened by the introductions of non-native species. I would like to see the red squirrel re-established over the whole of its former range. But it's not quite as simple as red squirrel, good; grey squirrel, bad. Attitudes have changed over the years, and while it is almost sanctified now, historically the red squirrel was blamed for damage to forestry and for this reason was mercilessly persecuted up until the mid-twentieth century.[15] A combination of habitat loss and persecution meant that the red squirrel in Scotland had to be sustained by multiple reintroductions to the extent that their current genetic make-up is primarily Scandinavian.[16]

Although we've uncovered a number of the differences in attitudes to animals and partially explained how they have developed, this is far from the complete picture. How attitudes develop and change over time is complex, and it cannot all be easily explained. These differences translate into glaring inconsistencies in what we are prepared to tolerate. And some of these inconsistencies have worked their way into the legislation that regulates how we treat animals.

The law to protect animals

Animal protection law has developed steadily since the first Acts in Britain were implemented in the mid to late nineteenth century. The first comprehensive animal protection law was the Protection of Animals Act 1911. The definition of animal was much more widely drawn than hitherto and it set out to protect animals from acts of commission ('Thou shall not … beat, etc.'). The Act specifically excluded animals that were not captive and thus free-living wildlife.

It was not until the 1960s that the tide turned towards legislation that was not simply drafted with a view to preventing harm and deliberate cruelty but towards a recognition of the needs of animals, that is, more towards offences of omission ('Thou shalt provide … food, water, shelter').

Much of this change in thinking came about in the wake of the publication in 1964 of Ruth Harrison's influential book, *Animal Machines.*[17] Harrison's book described new agricultural systems including battery cages for hens, individual crates for veal calves and tether stalls for sows. These are systems, she argued, that strip away the individuality of the animals themselves and turn them into mere production units. The book includes the memorable sentence:

> If one person is unkind to an animal it is considered to be cruelty, but if a lot of people are unkind to animals (especially in the name of profit) the cruelty is condoned and will be defended by otherwise intelligent people.

In 1985, much of my job as a newly appointed veterinary officer consisted of ensuring that farmers and others complied with legislation designed to protect the health and welfare of farmed animals. The Ministry of Agriculture's State Veterinary Service was keen to see high standards. But it became clear that there was truth in Harrison's accusation. Enforcement of the animal welfare legislation, even to the extent of prosecuting farmers, was far more likely to be directed at the small, inadequate and poorly resourced farms where the owner, struggling to make a living, was neglecting his animals, than at the larger, more highly mechanised and intensive operations. Perhaps it was because the former was more obvious whereas the larger farms operated behind closed doors and with the tacit protection afforded by near-universal use of systems such as sow crates and battery cages. Of course, action was taken when animals were neglected – but the ubiquity of the intensive systems and the fact that the law did not specifically preclude them meant they could not be challenged despite the mounting evidence of the harms caused. We'll see later how this argument remains relevant to much of today's exploitation of animals.

There is legislation covering all farmed animals, backed up by useful Codes of Welfare Practice, one for each species or class of animal – but considerable areas are only lightly regulated. Although the law imposes a duty of care on the keepers of all captive animals, not all species have equally detailed husbandry requirements. For example, there are few specific requirements for sheep and adult cattle. In contrast, there are detailed rules for laying hens, broilers and pigs to the extent that certain formerly widespread husbandry practices are now prohibited – that is, unfurnished cages for laying hens and the permanent tethering and crating of sows. Overall, while there is no requirement for farm premises to be licensed, there is a requirement for the person responsible for the animals to have sufficient knowledge of their care and needs – but there is no formal accreditation or third-party scrutiny.

In contrast, the use of animals in research is subject to probably the most comprehensive animal welfare legislation ever enacted. Almost every aspect of any intervention that takes place is regulated. And, in sharp contrast to farms, each premises and each operative is individually licensed and obliged to comply with a set of detailed conditions.

Legislation to protect wild animals is somewhat different. First, the vast majority is proscriptive – the prohibition of certain types of traps, for instance. Second, it is not comprehensive. Many species are barely mentioned, some not at all. Third, and perhaps most important, there are few circumstances where the operative has to demonstrate competence in any of the activities that have the potential to affect animal welfare. An exception is the use of snares in Scotland.

These differences are important. Similar species of animals, sometimes the same species of animal, enjoy greater legal protection in some circumstances than in others. The different circumstances have not developed by accident – they developed because of the way we keep, use and exploit animals. This is not supported by any ethical or scientific rationale. The differences are simply a reflection of our values.

The Five Freedoms

The furore that followed the publication of Harrison's book is said to have prompted the British government to appoint a committee chaired by the zoologist Professor Francis William Rogers Brambell to investigate the welfare of farm animals, and in 1965 the Brambell Report was published.[18] The report, a mere 84 pages long and masterfully concise, is perhaps the most influential animal welfare report of all time. Its recommendations eventually led to the formulation of the 'Five Freedoms', which now form the basis of much of the animal welfare legislation and practice in Britain and across the globe.[19] The Five Freedoms are:

1 Freedom from hunger or thirst – by ready access to fresh water and a diet to maintain full health and vigour
2 Freedom from discomfort – by providing an appropriate environment including shelter and a comfortable resting area
3 Freedom from pain, injury or disease – by prevention or rapid diagnosis and treatment
4 Freedom to express (most) normal behaviour – by providing sufficient space, proper facilities and company of the animal's own kind
5 Freedom from fear and distress – by ensuring conditions and treatment which avoid mental suffering

Looking at this list in the first quarter of the twenty-first century, it is hard to believe that any of it needed to be written down at all. Wasn't this the creed that all animal keepers routinely followed?

Evidently it wasn't, as legislation and Codes of Practice began to appear from the late 1960s onwards setting out mandatory conditions and recommendations. Over time, particularly since the European Union became involved, legislation has become more and more detailed to the extent that space allowances for certain species in certain circumstances are mandatory,[20] and, again for some species, there are rules setting out how that space is furnished.[21]

Despite the large body of animal protection legislation, there remain some strange and odd gaps in legal protection and the non-legislative norms which we apply. Some are historical, from a time when evidence was scant and attitudes were different. For example, tail docking of cattle and horses is illegal but it is legal to dock those of pigs and lambs. There is enough evidence from these species' anatomy and physiology to convince me that there is little difference between the horse's sentience and capacity to suffer and that of a pig. So why the difference? It's unclear, but it is probably something to do with differences in the way people perceive the capacity of animals to suffer. But it is most likely related to the value we ascribe to them. Perhaps, when the legislation was first proposed, as might be expected, more people voiced concern about the welfare of the horse than the pig.

Some differences in animal protection appear simply because the law hasn't kept up with the science. There is ample evidence, for instance, that decapods (lobsters, crabs, etc.) are sentient and quickly learn to avoid an unpleasant or aversive stimulus. However, in Britain decapods are excluded from the general animal protection law, allowing them be boiled alive.[22] Lobsters may be ordered online from some large retailers and are delivered in the post, alive, in a cardboard tube. One wonders how long a company offering a similar service for puppies would survive.

The Five Domains

In recent years animal welfare scientists have proposed a reformulation of the Five Freedoms into what are generally known as the Five Domains of potential welfare compromise (see diagram below). These address weaknesses in the older list by distinguishing between the physical/functional and the mental factors that contribute to an animal's welfare, reflecting an acceptance that the mental experiences of animals are important, and highlighting that what the animal itself experiences represents its welfare status.[23]

So what does this mean in practice? It means that simply meeting the physical needs of an animal – such as supplying sufficient food and water, keeping it free of discomfort, pain and disease – is not sufficient. It means that good welfare involves taking into account the animal's mental experience when meeting its needs. This includes considering whether animals have behavioural needs and to what extent meeting these needs contributes to achieving a positive mental state; this is where it gets complicated and controversial.

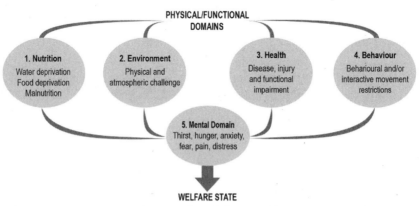

The Five Domains Model. Reproduced from Mellor, D.J., Beausoleil, N.J., Littlewood, K.E. *et al.* (2020) *Animals* 10: 1870. https://doi.org/10.3390/ani10101870. Licensed under CC BY 4.0.

The behavioural needs of animals

It is obvious that there are substantial differences between species of animals. The most obvious differences are physical, but there are often quite marked physiological differences between species which have a bearing on their husbandry requirements. Importantly, there are substantial differences in ecology and behaviour between species; a carnivore such as a lynx is generally solitary, while herbivores like antelopes generally live in groups and have complex societies.

Diet also has a considerable influence on behaviour, with carnivores generally perhaps only needing to eat once every day or so where a herbivore might spend 8–10 hours per day grazing and chewing the cud. What does this mean for animal welfare?

There are two terms frequently used by animal behaviour and welfare scientists when discussing behaviour: appetitive behaviour and consummatory behaviour. For example, appetitive behaviour is when the animal is searching for food or hunting and consummatory behaviour is when the animal has found the food and is eating it. Similarly, the appetitive behaviour of a rutting red deer stag might be rounding up receptive females to create a harem. Consummatory behaviour would be mating with the hinds.

For the most part, farmed animals and other animals in captivity don't have to bother with appetitive behaviour. The food is put in front of them, saving them the bother of having to look for it. The sow that might spend several hours a day foraging, turning over turf searching for grubs and roots, is simply provided with that day's nutritional needs in the form of a carefully formulated and measured amount. All it has to do is stick its head in the trough and eat.

But what if denying the sow the opportunity to exhibit its normal appetitive behaviour is the cause of distress and becomes a welfare problem? The sow might be physically comfortable, well fed and watered, as well as free from fear and disease, but what if the sow has an instinctive 'need' to forage and root in the ground? A barren environment with, for instance, smooth concrete floors with no manipulable material might exacerbate the problem. This is important in considering the mental factors that contribute to an animal's welfare state.

It has been hypothesised that the stereotypical behaviour that frequently develops in tethered and crated sows is a coping mechanism in response to confinement and the thwarting of appetitive behaviour. Sows which are not confined and have the ability to range rarely develop stereotypies.

What are the implications of this for the welfare of farmed animals, and for other captive animals? If animals, or at least some of the animals we exploit, have behavioural needs and these are being thwarted primarily by the confines of the environments in which we keep them, then animal welfare is potentially badly compromised. In the next few chapters we will see how exactly this occurs in different circumstances.

Differences in behaviour are adaptive

Like the physical differences between species of animal, the behavioural repertoire of species has evolved to allow the individual to cope better with its environment. This includes the behaviours that have evolved, at least in part, in response to the other members of the species with which the individual associates. In other words, the other sheep in the flock or the other wolves in the pack. Or, in the case of many carnivores, to a largely solitary existence. The differences in behaviour that we observe have a bearing on how we perceive the animal and, in particular, how the animal reacts to stressful circumstances. Let's consider a few examples.

It became obvious to me in my first few months of general practice that the recuperating cat was a much easier patient to manage than the recuperating dog. That is a generalisation, but in my experience a cat recovering from a fractured femur is inclined to hide away and rest whereas a dog with a similar injury will want to socialise and remain part of the 'pack'. It is very much more difficult to get the dog to rest. I believe this can be explained in part by the differences in canine and feline 'society'.

The ancestor of the domestic dog is the wolf. Wolves are social animals that live in packs. A wolf that cannot keep up with the pack is useless and risks abandonment and death. This behaviour is seen (and exploited by us) in our relationships with our pet dogs. A dog is desperately keen to please and remain part of the 'pack' – which we love – and expresses its fear of separation by barking and chewing the sofa – which we hate. In contrast, a cat, being primarily a solitary species, appears not to care about other cats or people. A cat with a broken leg has no pack to keep up with and hence rest is the preferred strategy. A dog, concerned about the loss of its position in the pack, will strive to keep up, even if that may damage the chances of the fracture healing well.

If the differences in the way cats and dogs appear to respond to pain and enforced rest can be explained by the differences in their respective social structures and how they behave as a consequence, is there a difference in the pain the dog experiences in comparison to that of the cat? It is impossible to say, but it would be unwise to assume that differences in observed behaviour are evidence that the dog is experiencing less pain.

Flocks and herds of herbivores develop for a number of reasons – for instance as an aid to finding food, during migration, the development of

36

harems with a single male defending a group of females prior to mating, and to exploit the benefits from safety in numbers. When a predator is on the prowl, it will look through the flock for the easiest pickings – the weak, the diseased and the injured. The 'stotting' or 'pronking' behaviour of deer, antelope and occasionally sheep – where, on being stalked and chased, the fittest animals jump high in the air while fleeing – has evolved to demonstrate that it is not worth chasing high-flying individuals. In contrast, the flock member with a broken leg or other injury will do its best not to stand out from the rest. It won't bellow or lag behind, and it will desperately strive to keep up. Anyone who has tried to pick out and catch a ewe with a broken leg from a flock will know exactly what I mean. Perhaps the stotting animal makes the lame sheep or deer that bit more obvious by drawing the attention of the predator to any individual that is unable to jump.

Does the sheep or deer with the broken leg that is striving to keep up with the flock and not draw attention to itself experience less pain than the dog with a similar injury? Again, it is impossible to say. And again, it would be unwise to assume that differences in observed behaviour are evidence that the injured sheep or deer experiences less pain.

It might be argued that farmed animals have adapted themselves to the conditions in which they are kept. It is true that generations of selective breeding have rendered farmed animals much more tolerant of the presence of humans, but this should not be taken as evidence that they have lost the entire repertoire of behaviours that helped sustain them in the wild. On the contrary, as we will see later, there is good evidence that, given the opportunity, farmed animals rapidly revert to the patterns of behaviour of their ancestors.

Do these differences in observed behaviour influence the way in which we perceive animals, and hence how we are inclined to exploit them? It's hard to say, but I believe it does. Take the law about mutilations – anaesthesia is required for the castration of dogs and cats at all ages and with no exceptions. But you can castrate a sheep up to three months of age without an anaesthetic.[24] There's no evidence that sheep feel pain less than cats and dogs, and so one is forced to conclude that we care more about cats and dogs than we do about sheep.

How does one counter that? As you might expect, it's not simple. It is certainly not as simple as just changing the legislation. Granted, the law as it relates to animals has developed piecemeal and is correspondingly complex. It is out of date, fails to take account of the latest science, in many cases fails to hold people to account for their actions, and has too many areas where responsibilities are unclear or absent altogether. However, the law and practice should reflect society's values. People's views and attitudes change over time, as does their appetite for change. Hence, the calls for change in the way in which egg-laying hens were kept pre-dated the evidence but resulted in changes to policy and legislation (see below).

Efforts to ensure compliance differ wildly

As efforts to improve animal welfare have evolved over time, differences between regulation and efforts to achieve compliance have developed. There is little point in enacting legislation unless there is compliance – and, as you will be expecting by now, when it comes to ensuring compliance with the legislation there are considerable differences between farmed, research and wild animals. Some of these differences are stark.

There are around 80,000 holdings registered to keep farmed animals in Britain. Given the dynamism of farming, the fact that more than one species of animal is kept on many farms, and that very small farms are excluded from the census, it is difficult to obtain a more accurate figure.[25] It is also difficult to calculate the animal welfare inspection effort of the two enforcement authorities (that is, local authorities and the Animal and Plant Health Agency), since both groups of officials have numerous other duties as well as seeking to ensure compliance with farmed animal welfare rules. A generous estimate would be around 100 full-time equivalents, giving one inspector for every 800 farms. Compare that to the Home Office inspectorate, the body charged with the licensing of premises involved in and ensuring compliance with animal research legislation. There are 153 premises licensed under the Animals (Scientific Procedures) Act 1986 (ASPA) and around 20 inspectors (full-time equivalents), giving around one inspector for every eight premises.[26]

Comparing the effectiveness of inspectorates is difficult, but in crude terms each Home Office inspector is responsible for the inspection of 100 times fewer premises than an equivalent inspector of agricultural premises. The job is different and the risks are different but even allowing for this it is beyond doubt that a great deal more effort is expended in ensuring compliance with the animal research rules than with those that apply to farmed animals.

When it comes to wildlife the differences are even more stark. There is no inspectorate devoted to the welfare of wildlife, and although licences to trap and kill wild animals are issued (in some circumstances) by the regulators (Natural England, NatureScot and Natural Resources Wales) there appears to be no routine inspection or monitoring of compliance with these licences. Certain activities which involve killing wildlife either need no licence or a general licence exists which allows the landowner to authorise anyone to trap and kill listed species.

There appears to be no obvious rationale for these differences. There are different regulators for farmed animals, research animals and wildlife, and the enforcement policies of each appear to have developed independently. But take into account what we know about animal sentience, irrespective of the use to which the animals are put, and by any objective analysis this is wrong.

Should the ecology of the ancestral form influence how captive animals are treated?

Scientists have long been aware of differences in animal breeding and rearing strategies. Many of the species we exploit have widely different strategies – compare the cow with its single calf which takes 2–3 years to become sexually mature to the chicken which begins to lay at five months of age and produces more than a dozen eggs (and in intensive units over 300 in a year). The number and survivability of their offspring raises an important moral question.

It is helpful to consider the breeding strategies of two species of animal, the frog and the elephant. From the age of two years, the common frog will produce several hundred eggs every spring. Neither parent cares for the hatching tadpoles. Of the eggs perhaps only two or three will make the transition from egg to tadpole to froglet to adult frog, thereby ensuring replacement. The rest die in one way or another. In contrast, the African elephant starts to breed at 10–12 years of age and has a gestation period of 22 months. Typically, the calf will not be weaned until it is 5–6 years of age. Over a 50-year lifespan a female elephant may produce six calves, that is, one every 6–7 years. Infant mortality is low.

These two examples are close to the extreme ends of a continuum. Scientists have described these as K- and r-selection strategies, and although the theory has been somewhat discredited and revised since it was first described in the 1970s, it is still a useful way of articulating differences in the way species breed. The table below summarises the differences between the two strategies. In reality, most organisms typically demonstrate an intermediate strategy somewhere along the continuum.

Summary of characteristics of r- and K-selection strategies

r-selection	K-selection
r-selected species have a high growth rate but low survivability ('cheap' offspring)	K-selected species have a low growth rate but high survivability ('expensive' offspring)
The body size of offspring is typically small and they have an early onset of maturity	The body size of offspring is typically larger and they have a late onset of maturity
There are usually many offspring per brood and often little or no parental care (resulting in high mortality)	There are usually very few offspring per brood, each requiring high levels of parental care (resulting in low mortality)

At this point you would be forgiven for asking 'What has this got to do with the way we treat animals?' After all, this is the way that certain species have evolved to adapt to their environment. If some species produce a large number of offspring and invest little care in them, why is that relevant to discussions about animal welfare? When it comes to common frogs, perhaps the answer is that it doesn't a great deal. But we are considering captive animals, among others, which we exploit for own purposes – and some of them produce considerable numbers of offspring.

For example, the red jungle fowl, the ancestral form of the domestic fowl or chicken, lays between 10 and 20 eggs per year in two clutches. Perhaps only four or five of these will fledge, the rest dying from various causes. These losses are not as extreme as those of the frog. The red jungle fowl has evolved an intermediate selection strategy but, as a species, it clearly copes with considerable mortality in its offspring. But what about the domestic form? Decades of selective breeding mean that broiler breeder hens (these are the birds that lay the eggs that hatch into the birds we rear for meat) produce 200 or so eggs per year. Layer breeder hens (the birds that lay eggs that hatch into the birds that eventually lay the eggs we eat) produce similar numbers. Typically, the mortality of the hatching chicks is substantially lower than in the ancestral form (although note that most male chicks from laying hen strains are killed when a day old).

Similar comparisons can be made between the domestic pig and its ancestral form, the wild boar. In general, the wild boar will produce 4–8 young in a litter, of which perhaps two or three will reach weaning age. In contrast, the genetically 'improved' domestic pig will produce 10–12 piglets per litter, of which, with good husbandry, 95% will reach slaughter weight. Again, mortality in the domestic pig is considerably reduced in comparison to the wild form.

And yet, despite this difference, leaving aside the obvious commercial benefit of further reducing mortality in farmed animals, the pressure from the welfare lobby is to cut the death rate to as close to zero as possible. Is this justified, given that, in the wild, few red jungle fowl and few wild boar will ever reach maturity? Isn't a certain level of mortality in farmed chicks and piglets to be expected, perhaps even to be tolerated, given how far both species are along the spectrum towards an r-selection strategy? It's not as if we are dealing with cows and calves, which, in comparison, sit quite close to the elephant on the r–K scale.

It's a difficult question to answer. I once worked with a veterinarian who considered animal welfare as a hindrance rather than an obligation for animal keepers. But then he said something which has stayed with me: 'As soon as you put a fence around them, then they become your responsibility.' It's a compelling argument. It's not a new sentiment but it has resonance and currency simply because it is now the basis of much animal welfare policy, where the duty of care has become a legal obligation. Does the duty of care mean ensuring that mortality of the growing broiler chick or the unweaned piglet must be reduced to zero? Particularly when it is well-nigh impossible to achieve. This an ethical issue. Farmed animals are far removed from their wild antecedents and their provenance. We have bred these animals to reproduce in large numbers and to grow quickly. It is incumbent upon us to deal with the associated consequences. These animals are now under our care, and striving to keep them healthy is both a moral and a legal imperative.

It might help to consider the conundrum in the round using examples from other groups of animals. Huge efforts are expended to protect the welfare of research animals, with an obligation to report each and every unexpected death to the authorities. There is no similar obligation for farmed animals. This applies even to zebrafish, a species increasingly used in animal research. When it comes to breeding the zebrafish is quite close to the common frog in its position on the r–K scale. The same approach applies – strive to keep them healthy regardless of the scale at which they reproduce.

The argument that we should intervene against certain predatory animals such as badgers and sparrowhawks because of the impact they have on the populations of songbirds like warblers and thrushes is not so far removed. We will see in Chapter 7 that the presence of sparrowhawks has no effect on bird populations. Sure, they kill and eat a variety of small birds. They have evolved to do that. But similarly, songbirds have evolved to produce a surplus which ensures, other things being equal, that the population is sustained in the presence of sparrowhawks and other predators. But, given that we strive to protect piglets and chicks from an early death, shouldn't we do the same for the birds in our gardens and hedgerows?

I don't believe we should. Once we take control over an animal, we assume moral responsibility for its welfare. Wildlife is different and, assuming that human intervention is carefully controlled and limited, does not need a helping hand, either to protect the welfare of songbirds or to routinely manipulate populations.

Why care about the welfare of captive animals and ignore the suffering of wildlife?

Given that a fox does not take into account animal welfare when it catches, kills and eats a rabbit, are we obliged to take into account that an increase in the population of a predator might be considered detrimental to the welfare of its prey?

A criticism from those sceptical of the value of proposed reintroductions to Britain of formerly native predatory species such as lynx, bear and wolf is that the harm done to the welfare of prey species would be unacceptable. Why, it is asked, do you object to the killing of foxes by organised hunts on grounds of animal welfare when you advocate the introduction of predators which will kill animals in ways that inevitably cause suffering? It's a case of double standards, isn't it?

It is a good question, but not one that science can answer. This is a matter of personal values and ethics.

Wildlife is not cruel. Wildlife is neither kind nor humane. Wildlife just is. That is not to say that a deer tracked, caught, killed and eaten by a lynx hasn't suffered. Nor would I deny the suffering of the fox set upon by a pack of hounds. But it's different. The first happens, whether we like it or not. The second is a matter of choice. Our choice. We exercise choice as moral beings – either individually or collectively. I choose not to support inhumane and lethal wildlife interventions such as fox hunting because this is part of my values. I'll go further. I neither wish to intervene to protect one species from another (except in the short term when conservation priorities demand it) nor seek to change the experience of wildlife. But I'd choose to have lynx reintroduced to Britain because I believe in enhancing biodiversity and the re-establishment of effective predator–prey relationships.

How can we achieve consistency?

Society exploits animals inconsistently. We've seen how these differences develop in how we treat animals and the reasons why it varies so much: history and tradition, gaps in science and evidence, variations in inspection and enforcement effort, cultural mores and the fact that legislation and policy lags behind society's changing values.

I advocate two main ways to effect change. The first involves making better decisions. To do this, we need to start with better information – which is one of the main reasons for writing this book. The next step is to take that information and use it to make decisions about where the boundaries in our relationships with animals lie. That is, where you and I, as consumers, draw the line in animal exploitation.

But because consumer choice won't solve everything, a second approach is needed – the involvement of wider society. Animal welfare science doesn't exist in a vacuum. Evidence is important but so is society and its values. Although the evidence shows that the now outlawed barren battery cages are detrimental to the welfare of the laying hen, that evidence didn't come out of the blue. It wasn't science that drove the change to colony systems with more space, dust baths, perches and nest boxes. Granted, it was science that demonstrated that chickens would choose to 'work' for less barren environments but, in the first instance, it was citizens who called for change, saying they would no longer tolerate hens being kept in these conditions. It was then and only then that the science began to investigate the hens' needs, paving the way for the more spacious, furnished colonies that are now found on poultry farms throughout Europe.

The rest of this book sets out the way in which the main arbitrary groupings of animals are exploited and how that exploitation is regulated. This is not a book about animal husbandry, and the detail is deliberately skewed towards animal welfare considerations rather than serving as a 'how to' guide. Because these animals are not kept in a vacuum, and because we are in the midst of a climate and biodiversity emergency, I will touch on both these areas of concern when dealing with each of the groupings.

3

The Welfare of Farmed Animals: an Overview

The farming of animals in Britain is primarily a commercial enterprise which involves many thousands of people aiming to provide good-quality, safe food for many millions of people. The welfare of these animals is important, or should be important, to producer and consumer alike. And to be clear, with the exception of a tiny minority, the welfare of animals in the care of farmers is and always has been important. Farmers take pride in their animals. This is not simply a matter of profit, although that is an important motivator. In short, in my experience, most farmers care about the welfare of their animals.

The farming of a number of species of animal forms a substantial part of the UK's economy; the annual output over the last several years is around £15 billion.[1] The products derived from those animals, whether meat, milk, leather or other animal products, form an important part of our daily lives and diet: 86% of the population currently eat meat in their diets.

Over 70% of Britain's land is given over to agriculture, and a high proportion of that total is taken up either by the grazing of farmed animals or the production of the food that they eat. One way or another, farming is an important part of where we live.

Different experiences, different attitudes

I've never farmed animals. As a schoolboy, and later as a student, I spent several lambing seasons on sheep farms in Wales and Scotland and summers working on farms with beef suckler herds in the west of Scotland. For a few years I kept a small number of sheep on land I rented in Somerset. I got my hands dirty and my knees bruised but I would never call myself a farmer. For a start, I wasn't making a living from it. I have nothing but admiration for those who do. It's one of the toughest gigs of the lot – unremittingly hard work over long hours and often for poor rewards. Particularly in the uplands, rearing farmed animals is isolating and precarious.

While working as a veterinarian, I came to realise that farmers could be starved of company. Take, say, a mixed cattle and sheep farm in rural Aberdeenshire. In the short days in the depths of winter, the farmer might not see another soul for days. I might visit to test a herd of cattle, and once we had finished the job, over tea and cake the conversation would range far and wide – from the price of cattle to emerging disease risks to the future of farming. I learned a lot from those chats. Sometimes the conversation would stray into the attitudes of the public to farming – which was almost universally considered to be unsympathetic. Farmers complained that people didn't understand farming and were ungrateful despite the availability of abundant, cheap food. My urban-dwelling friends were equally bemused (and ignorant) about farming and considered farmers to be out of touch, serial grumblers and always on the lookout for another government handout.

None of this is true – on either side of the argument – but it is indicative of the vast difference between the experiences of farmers and consumers: it struck me then, as it does now, that there is a huge gulf in understanding and a mutual lack of empathy. And yet the relationship is symbiotic – without farmers there is no food, and without people there is no market for the food.

One area where the gulf is at its widest is in the appreciation, or lack of appreciation, of the way animals are farmed. The welfare of farm animals is an enduring concern and a divisive and controversial subject. Some people take an ethical position and argue that the keeping and rearing of animals for food is inherently wrong, while others campaign for improvement in the way animals are husbanded. Many others appear indifferent. In response, farming representative bodies and individual farm enterprises seek to bolster and maintain a good image of their businesses.

The attitude of the public to farmed animal welfare has been well studied. For example, studies show that the majority of the public are concerned about animal welfare in modern production systems. However, a substantial number of people adopted dissonance strategies to enable guilt-free meat consumption. In plain words, this means that a subgroup of consumers has not let their concerns about animal welfare affect their consumption of animal products.[2] There is evidence that people, while expressing concern about farmed animal welfare, take many factors into account in their food purchasing decisions. Many consumers express a willingness to pay more for meat from high-welfare systems, but willingness to pay a premium has its limits and must be linked to other quality characteristics.[3]

This book argues for a more consistent approach to the welfare of animals, one which takes account of sentience rather than simply the arbitrary categories into which we squeeze them. This applies as much to farmed animals as it does to any other type of animal. But with the possible exception of pets, for most people their main interaction with animals is an indirect one – that is, the consumption of animal products that are several steps removed from the rearing and the slaughter of the animal. It follows that to make an informed choice about which animal products you choose to consume, you need good-quality, unbiased information. And this is far from easy to come by.

Husbandry of farmed animals

The farming of animals in the UK consists of a wide range of systems. At one end of this range is the system of rearing ewes and lambs in the hills and uplands. Outwardly, this hasn't changed much in well over a century and is an example of low-input, low-output farming generally described as 'extensive'. At the other extreme, modern broiler (meat) chicken production is an example of high-input, high-output farming and is generally described as 'intensive'. However, for each farmed species there are examples of both intensive and extensive husbandry.

While I contend that most keepers and owners of farmed animals care for their animals with a genuine commitment to their welfare, it is clear that when businesses are dealing with, for instance, hundreds of pigs or tens of thousands of chickens, the sort of care that would be applied to an individual animal when keeping, say, ten birds is simply not possible. This brings us back to Ruth Harrison and her compelling argument that when cruelty to animals becomes the norm, the associated husbandry systems are defended by intelligent people.[4]

As we've seen, Harrison's book and the changes to government policy that followed have had a profound effect on how we deal with farmed animals. But given the staggering increase in the scale of animal production in Britain and elsewhere since *Animal Machines* was first published in 1964, it is important that all systems of production are open to scrutiny and challenge, and that where there is evidence of welfare problems, we make appropriate changes. This is where society comes in. We have to face up to the need to make decisions about what is acceptable – either via legislation or by acting as consumers, individually and collectively.

Farms are, with few exceptions, businesses. There are farms in Britain, with or without animals, that are run simply to provide a subsistence for the occupants, but these are few in number. Like the rest of us, the majority of farmers want to make a decent living, and to provide for their families and for their retirement. That means that the farm and the animals kept there have to be profitable.

Just like any other business, a farm is subject to competition and the effects of the economy of scale. For example, when I first qualified in the late 1970s the average size of a dairy herd in England was around 30 cows. By 2020 the average size of herd was 155 cows.[5] At the same time, the amount of human labour required on dairy farms has fallen sharply. One to two people can now care for a 250-cow dairy herd. This economy of scale has been facilitated by the mechanisation of many of the necessary daily tasks. The situation is no different on most other farms; almost without exception, as the numbers of farmed animals per farm have increased the number of stockpersons per farm has fallen. With many more charges per person, does this mean the animals are less well cared for?

Many people would conclude that an increase in the number of animals that a person cares for will inevitably mean a lower standard of care, but it is not as simple as that. Farmers are better educated and better equipped than in the 1970s. Nutrition, housing, labour-saving equipment, and disease control and treatment are all on a higher plane. All of which means that more animals can be safely cared for by one person. However, while mechanisation and improved practices play an important part in animal care, the steady reduction in the amount of skilled labour on farms remains a concern; there has to be a limit as to how many animals each person can be responsible for.

There is another important relationship relevant to farmed animals, particularly those kept in intensive systems, and that is the relationship between stocking density and production. The figure below shows a schematic representation of the relationship between stocking density and output per unit area. The output may be in kilograms of meat produced per square metre, litres of milk per hectare, eggs per cage, or whatever. Stocking density can be measured as the number of animals in a given area. The graph shows a direct linear relationship between stocking density and output per unit area until an inflexion point is reached (solid line). Thereafter output per unit area remains steady despite the continued increase in stocking density. One can postulate a number of reasons why output stops rising beyond a certain stocking density. For instance, the number of animals in a given area reaches a point where systems to maintain a suitable environment can no longer cope, or food and water supplies are barely adequate, etc. Irrespective of the reasons, it is clear that there is a limit to production, and although that limit will vary from system to system and from species to species, each unit will be similarly constrained. However, if stocking density reaches a point where disease can no longer be controlled or environmental controls begin to fail, then output per unit area will fall (dotted line).

As we will see, a combination of public concern and scientific studies of the relationship between stocking densities, the quality of that space and animal welfare has had a profound effect on the regulatory standards that now prevail for broiler chickens and egg-laying hens. Similar discussions and outcomes have influenced the keeping of pigs, although to a lesser extent

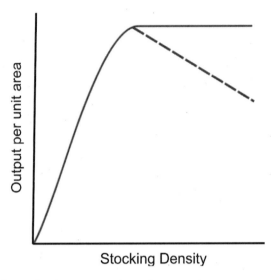

The relationship between stocking density and output per unit area

than for poultry. Fewer changes have been imposed on the conditions for the less intensively reared species like cattle and sheep.

Although the vast majority of farmers would be horrified if they were accused of stocking their units to the maximum density that they could get away with (that is, at the inflexion point and beyond, as shown in the graph above), it is an inescapable fact that various factors have pushed farmers to stock at or near this point. Factors that make achieving a decent return difficult include the rising costs of inputs, price pressure from retailers and competition, among many others.

There are a number of issues that may affect the welfare of all the main species of farmed animals. I touch on each of these here and consider them in more detail in Chapters 4 and 5.

The impact of farming on natural behaviour

The ancestral and undomesticated forms of pigs, cattle, sheep and chickens are social animals, living in dynamic groups which contain individuals of both sexes and all ages. Social structures allow most individuals to make the best of their environment, keep safe, find food and breed success-fully. This includes foraging, territorial defence, courtship and mating, nest building and rearing young. Much behaviour is innate – the animal is 'pre-programmed' – although the immediate environment and experience influences how behaviours are expressed. And obviously, learning plays an important part too.

The very nature of farming has a substantial impact on animal behaviour. As soon as a fence encloses a group of animals, their behaviour changes. Instead of dynamic social groups, farmed animals are now kept in groups the make-up of which is decided by the farmer. There is evidence that increasing group size, particularly above the size that would form naturally, may affect the expression of the full repertoire of normal behaviour.

Until relatively recently it was believed that selective breeding had virtually eliminated these behaviours. The full repertoire was no longer needed as the farmer now provides food and shelter and makes decisions about when and with whom to breed. Why select to retain the protracted foraging behaviour of the pig when on the farm the food is directly placed in front of it? However, these innate behaviours are more resilient than we first thought. Studies in the 1980s showed that even with no previous access to the outdoors, it only took a few weeks for pigs to adopt a behavioural repertoire similar to that of the European wild boar in wild or semi-wild conditions.[6] Social groups formed as before and roots were grubbed up from the soil as if generations of artificial selection had never occurred. While not conclusive, this suggests that the full behavioural repertoire of domestic animals is retained despite generations of selective breeding.

In some cases, the natural variation in a particular behavioural trait has been exploited when selecting animals for production. A hen will often become 'broody' once it has laid a clutch of eggs. It will then sit firmly over its eggs, and when people approach or try to remove the eggs, it will threaten the person by erecting its feathers, emitting a warning – a characteristic drawn-out, throaty squawk – and pecking aggressively. Broodiness develops as the last egg in the clutch is laid and is therefore not a desirable trait in the laying hen. Producers want eggs laid continuously, and because the tendency for a hen to go broody varies from breed to breed the modern laying hen has been selected from those breeds less likely to exhibit this behaviour. This does not mean the trait has been lost, just that it has been heavily selected against, so as not to be expressed.

Despite recent legislation which changes the way some animals are kept, such as the provision of furnished cages for laying hens, the behavioural repertoire of some farm animals remains curtailed, generally because of the barren environment in which they are housed, particularly in intensive systems. On the other hand, most sheep and beef systems allow free ranging, and ewes and cows keep and suckle their progeny for something approaching a natural weaning period. Between these extremes there are many examples of systems of keeping farmed animals that inhibit natural behaviour to a greater or lesser extent. And as we saw in Chapter 2, the inability to express certain types of behaviour has been demonstrated to adversely affect animal welfare.

Disease

Despite enormous advances in veterinary medicine in the last 70 years, disease remains a major and enduring threat to the welfare of farmed animals (and the profitability of farms). Unchecked and untreated disease causes enormous suffering.

Expensive projects since the Second World War have reduced and eliminated many of the serious endemic and exotic diseases such as brucellosis, swine fever, Newcastle disease and foot-and-mouth disease, but the threat of re-incursion remains. And we continue to struggle to control tuberculosis in cattle and avian influenza in poultry. Add to that a range of insidious parasitic and infectious bacterial and viral diseases, and the farmer and his or her veterinarian have a continuing challenge to address. Thankfully, the availability of better medicines and vaccines, along with better trained and more knowledgeable farmers, supported by increasingly specialised veterinarians, provides the means to address most of the problems. But continued surveillance and vigilance is required.

Mutilations

A substantial proportion of farmed animals are mutilated in some way or another. That is, part of the animal's body is permanently removed, such as part or all of the tail, the horns, part of the beak or the genitals. Mutilation also includes any means of permanently identifying farmed animals using a method which causes tissue damage, such as applying an ear tag or a tattoo. The word 'mutilation' might seem a loaded term but it seems appropriate to avoid using a euphemism – these procedures involve cutting into living tissues.

Farmed animals are derived from species that, prior to domestication, evolved in the same way as other animals. That is, natural selection acted on the population so that the physical and behavioural traits that favour an individual's survival become widespread in the population. It follows that the horns, the tail, the beak and the testicles are not mere redundant appendages. They serve a purpose: communication, protection, selecting food and sexual reproduction, for instance. So it follows that the removal of these appendages by farmers (or, in the case of cats and dogs, by veterinarians on behalf of their owners) is for our convenience, not for the convenience of the animals involved. That is not to say that mutilations are done for no reason. In many cases, such as dehorning cattle and the beak trimming of poultry, it is to prevent the animals injuring each other. However, whatever the reasons behind mutilation, the tissues that are removed are part of the animal's body and are formed of sensitive tissues served with blood vessels and nerves. Their removal causes pain.

The mutilation of farmed animals is governed by detailed and compre-hensive legislation. The Mutilations (Permitted Procedures) (England) Regulations 2007 list 16 different permitted procedures for cattle, 13 each for pigs and sheep, and 12 for birds.[7] These procedures are governed by detailed rules which define the methods that may be used, set maximum ages at which the procedure can be done, and specify the qualifications of those performing the task. Some procedures may be done without anaesthesia provided the animal is not above a certain age. Some procedures, particularly the more invasive types, may only be performed by a veterinarian.

Among this long list there are several common mutilations which are so much part of the routine of farming that most of us don't give them a second thought. All male calves and most lambs, with the exception of those retained for breeding and a proportion reared in bull beef systems, are castrated. Most horned cattle are dehorned, or more commonly disbudded at an early age. A high proportion of lambs are tail-docked, and we may come across the tails and those distinctive rubber rings as we walk the fields in spring.

The law requires cattle sheep, pigs and goats to be permanently, and in most cases uniquely, identified. This is done by applying approved numbered ear tags using equipment which is rather like a scaled-up version of the machines used to pierce our own ears. Tail docking of pigs is an unusual case in that the procedure is prohibited except where circumstances require it to be done to prevent injury; it is, however, a common practice. Few pigs are castrated in Britain since the majority reach slaughter weight long before the males become sexually mature. The vast majority of laying hens have the tip of their beak trimmed, another procedure that the government and welfare organisations are seeking to phase out.

Some procedures are prohibited outright: for instance, the tail docking of cattle and the removal of antlers in velvet from farmed deer. It's worth noting that the mutilation of farmed animals in some countries is much more extensive and much less heavily regulated than in Britain. For example, in many parts of the USA there is no requirement to use analgesia of any type when dehorning cattle.

You might ask 'why mutilate?' There are many reasons, but in simple terms certain parts of the farmed animals' anatomy are routinely removed because the presence of these parts has been deemed incompatible with the systems in which they are kept. Horned cattle, when housed in loose-housing systems, are prone to gore each other, which can lead to severe injury. Uncastrated cattle and sheep when they reach puberty are inclined to fight, and mate with immature females. Pigs which have not been tail-docked are prone to outbreaks of tail biting, which can lead to serious injuries particularly when husbandry is suboptimal. Laying hens which have not been beak-trimmed that are kept in free-range and other non-cage systems may start feather pecking, which can lead to serious injury and cannibalism.

Do mutilations cause the animals involved to suffer? Given the knowledge we have about the anatomy and physiology of each of these species, it would be a reasonable assumption to conclude 'yes, they do'. Take the commonly used procedure of removing the growing horn buds from calves. There are numerous studies that show cautery disbudding and horn amputation stimulate definite pain-related behaviours during and after the procedure.[8]

Are mutilations necessary? That depends on who you talk to. On a simple, perhaps oversimplified analysis, one could conclude that they are not. If the pig industry can produce pigs without routine castration, then why not sheep and cattle? If cattle can be produced without having to dock their tails, why not pigs and sheep? If broiler chickens can be produced without routine beak trimming, then why not egg-laying hens? Why not use polled (naturally hornless) cattle, obviating the need to dehorn?

Of course, it is not as simple as this. While tradition undoubtedly plays a part and there are good reasons to question the continuance of these practices, there are good reasons to be cautious about prohibition without alternative means to manage the consequences. In the chapters on farmed animals that follow, I'll examine the case for the routine mutilation of each species individually.

Transportation and slaughter

It is an inescapable fact that farmed animals are produced for one purpose. That is so we can eat them or the eggs, milk or other products derived from them. Even dairy cows end up as meat in the final reckoning. Any consideration of the exploitation of farmed animals therefore has to consider the transportation and slaughter of animals. There is a huge body of science about the transport and slaughter of farmed animals: Are transported animals unduly stressed? How much weight do they lose in transit? What's the optimum resting period in journeys over 24 hours? What's the best method of humanely slaughtering a pig? In reality, although the subjects are complex, there are basic principles which, while few would argue with them, seem extraordinarily difficult to apply consistently.

To create meat, the animals have to be slaughtered, and to be slaughtered they have to be transported to the slaughterhouse. Enduring citizen concerns about the transport of animals and the methods of slaughter have spawned myriad scientific studies and a plethora of legislation. Despite the continuing controversy about both subjects and the majority of public opinion that is supportive of change, politicians find it difficult to subscribe to (or to translate into legislation) two basic principles that I believe should underpin the views of all thinking people:

- Farmed animals should be slaughtered as close to the point of production as possible.
- All slaughter should be done in a humane way by rendering the animal immediately insensible to pain.

You would have thought that adhering to both of these principles would be straightforward, and that there would be little argument about either. But it's not, and there still is a great deal of argument. Let's deal with each in turn.

Transport

With few exceptions, farmed animals are transported between two points for one of two reasons: for further production (that is, to be fattened for slaughter, to be bred or to be milked) or for slaughter. The reasons for animals being moved for further production are complex, but in Britain there are traditional seasonal movements and sales which have developed to allow for cattle and sheep to be moved from the uplands where grass becomes scarce in the winter to low-lying areas where supplies of feed are more readily available in the form of a longer grazing season, conserved grass (silage and hay) or cereals. These movements involve hundreds of thousands of cattle annually,[9] and similar numbers of sheep.

A very small proportion of these movements involves the transport of cattle and sheep to continental Europe for further production and slaughter, although the UK government has signalled an intention to prohibit this trade.[10]

The transport of animals to slaughter is altogether different and does not follow tradition. There are many fewer slaughterhouses in the UK than there were immediately after the Second World War. There has been a steady decline from a peak of some 5,500 in the mid-1950s to the current number of around 240. But despite this reduction there is plenty of capacity, and although there are often cries that there aren't enough slaughterhouses it is not the absolute number of premises nor their overall capacity that is the problem. Consolidation, specialisation and contract constraints often mean that accessing a nearby slaughtering service of the type required is far from easy. For example, there are fewer than 10 premises willing to slaughter cull cows, and over 95% of culled sows are slaughtered in one premises. This specialisation has come about because of the demands of processors and retailers who want a product that fits in with their supply logistics. Sow meat is very different from pork derived from a 20-week-old porker and has a different market. Obviously, this has implications for the more remotely located farmer, who has to have his culled sows transported many hundreds of miles – and for the sows that have to endure the journey.

The net effect is that the contracts the bigger companies have with supermarkets mean that your local slaughterhouse is unlikely to agree to process the six lambs the smallholder has spent the summer getting to prime condition. It simply doesn't make commercial sense to insert such a small number of animals into an already complex schedule of 2,000 lambs destined for one of the large retailers.

Capacity is being consolidated into fewer and larger premises. In England, 13 slaughterhouses are responsible for 56% of cattle slaughtered, while the smallest 74 premises are responsible for less than 1% of the total.[11] The reasons for this are complex but are not so different from the economic drivers which caused the demise of the village cobbler, baker and seamstress. Mechanisation and the falling price of transportation mean that small slaughterhouses, like small bakers and candlestick makers, struggle to make a living. All of this means that animals are being transported longer and longer distances for slaughter, and that gaining access to slaughter services for the small-scale operator is becoming increasingly difficult.

This does not mean that locally produced and slaughtered meat and poultry meat is impossible to source. My local butcher, deep in the West Country, sources lamb, beef and pork from local farms; the animals are slaughtered less than 20 kilometres from his shop. I trust this to be true because he tells me so. And I trust what he tells me because he knows what I used to do for a living. This comes at a price, however. The meat, while consistently high quality, is substantially more expensive than that from the supermarket around the corner.

There are regular calls for a network of smaller slaughterhouses to cater for the smaller producer supplying butchers and farmers' markets. There are proponents of mobile slaughterhouses. The idea is that a lorry-based facility will run from village to village providing a service for anyone who needs it. On the face of it, both notions are attractive since distances from farm to slaughter can be kept low. But it is difficult to see how either of these ideas can be put into practice without some sort of subsidy. In comparison to the costs of each animal slaughtered in one of the large premises that now dominate the sector, building and running small premises is expensive. And having spent much of my career involved in the regulation of the meat industry, I am dubious about how the notion that subsidies or substantial regulatory concessions for small premises would be received by politicians and economists. In any case, it is perverse for the taxpayer to subsidise the provision of facilities for the top end of the sector while the rest has to manage with none.

Slaughter

Eating meat involves animals being killed. Most of us choose to let others do the deed on our behalf in the belief that is done humanely. The law is clear and firm:

unless there is an exemption for religious reasons (see below), all animals must be stunned before you kill them. Once an animal has been stunned, it must remain unconscious and unable to feel pain until death. Once stunned it must be killed immediately. Stunning methods that instantly kill an animal can be used, that is, stunning and killing occur in one operation. Like so many of our interactions with animals, ensuring that stunning and slaughter are humane depends on the competence and commitment of the operatives. Training and supervision need to be of a high order. Fortunately, there are many legal safeguards:

- People stunning and killing animals must be demonstrably competent and have the appropriate licence.
- CCTV is mandatory in the parts of the slaughterhouse where animals are handled.
- There is mandatory veterinary supervision during the hours of slaughter (except for the very smallest premises).
- All premises are obliged to appoint an Animal Welfare Officer.
- Methods of stunning and slaughter are specified in law.

However, despite this, there are concerns about stunning and slaughter. First, there have been too many times when undercover filming shows poor practice and sometimes gross cruelty. Clearly, when this happens, it is almost invariably a failing of training and supervision, and the industry needs to be clear about how it is addressing this.

Second, there are concerns about some of the more widely used methods. The law sets out in some detail which methods may be used. There is good evidence that most systems are effective if properly maintained and applied. Despite that, there is evidence that some of the commonly used systems may not be effective in delivering a humane stun. For example, gas stunning of pigs is commonly used in large slaughterhouses. Although inert gases such as argon may be used in combination with or without carbon dioxide, carbon dioxide on its own is widely used. There is good evidence that carbon dioxide at the concentrations necessary to stun the pig is highly aversive, which means that pigs may suffer prior to becoming unconscious.[12] This was recognised by the European Union's science advisory committee in 2004,[13] but despite efforts no practical humane alternative has been devised, a sad fact that was acknowledged by the same committee in 2020 when it concluded that stunning of pigs using carbon dioxide was inhumane.[14]

Another area of concern is the slaughter of poultry. There are two main methods: electric water-bath stunning followed by exsanguination and gaseous stun/kill systems (controlled atmosphere killing, CAK). While the latter is widely used in larger slaughterhouses, the former is much more common in small to medium-sized premises,

Electric stunning involves movement through a constant-voltage, multiple-bird, electrical water-bath stun system. This system subjects live birds to stressful and painful shackling (while being held upside down by their legs), and the potential exists for them to receive pre-stun electric shocks and induction of seizures while still conscious. The existing electrical water-bath stunner settings are not necessarily based on sound scientific data, and research indicates that they are not effective in all birds. Further, in multiple-bird electrical water-bath systems, birds may miss the stunner completely. There is clear evidence that some birds may still be alive when they reach the scald vat. For these reasons, electrical water-bath systems are increasingly under scrutiny on animal welfare grounds. CAK systems stun birds while they are still in their transport crates, and thus avoid many of the welfare problems associated with the live-hang process and electrical water-bath stunning.[15] However, CAK systems are expensive to install, require considerable space and are impractical for smaller premises.

In order to satisfy the tenets of certain religious groups, there are controversial exemptions to the requirement to stun an animal prior to slaughter. The position of the British Veterinary Association (BVA) is clear and simple: 'All animals should be stunned before slaughter.'[16] This is an evidence-based position and, combined with the BVA's practical proposals to phase out non-stun slaughter, represents a mix of principle and pragmatism – a position with which I completely agree.

Summary

The systems that are used for the farming of animals are many and varied. It is clear that farmers care about the welfare of their animals but there are significant concerns about the ability of some farming system to ensure animal welfare. Add to that routine mutilations and the curtailment of natural behaviours, and one can see that the exploitation of farmed animals involves a number of practices that should cause people to pause for thought. The next two chapters delve in far more detail into the husbandry of the modern farmed animal. And we will see that there is a considerable difference between the best and the worst.

The two principles that I set out earlier – minimising transport times for animals and ensuring that animals are slaughtered humanely – are not being observed. Leadership, probably political leadership, would be required to effect beneficial changes, particularly as this would mean greater intervention in the market and more regulation. It is unlikely to come about by itself because of the way in which the sector and its supply chain has developed. Which is a shame, because I believe that without change to ensure that long-distance transport is minimised and that suffering at the time of slaughter is addressed, more and more people will vote with their feet and stop buying meat.

4

Grazing Animals: the Best, and Some of the Worst

In 1996 we moved to Somerset. I'd only ever been to the county once before but a new job seemed to offer the challenge I sought, along with the chance to escape the loathed Home Counties. We said yes, and within weeks we'd sold up in Surrey and were renting a tiny cottage at one end of West Sedgemoor, one of the many tracts of low-lying land that make up the Somerset Levels and Moors.

The job was tough – I was the new, inexperienced and untrusted boss charged with reorganising the operation. There wasn't open revolt, but neither was I received with open arms. Then, out of the blue, the biggest upset to the cattle sector for 50 years – a full-blown food crisis in the form of a link between bovine spongiform encephalopathy (BSE) and human health. Overnight, the job was turned upside down. Then my mother became terminally ill – at the other end of the country. My head began to spin. But then I discovered rural Somerset – and it began right outside my front door; here was the perfect antidote to the suburban sprawl we'd left behind and the rigours of work. Laid out in front of me were huge areas of low-lying grassland, prone to flooding in the winter and with nary a soul to be seen.

I set out to explore. Once the winter floods recede and the mud begins to dry, the Somerset Levels and Moors come to life. I opened up underused and overgrown footpaths, drinking in the wildlife: roe deer, badgers, buzzards, owls of several species and insects beyond number.

This hasn't happened by accident – although shaped by human hands with its drainage ditches and droves, the landscape is maintained, primarily, by grazing. These glorious meadows, full of drumming snipe, bubbling curlew and, more recently, a re-established population of cranes, simply wouldn't exist without cattle. Grazing of the type practised on West Sedgemoor by the RSPB keeps the vegetation short, the shrubs at bay and, with the right management in place, maintains a wide variety of plants: the very plants that support the insects that feed the birds. Of course, this type of conservation grazing is not the exclusive preserve of conservation bodies – it happens all over the country – but it should not be confused with more intensively managed, improved grassland where biodiversity is often much reduced.

I still walk the Somerset Levels and Moors today. Twenty-five years ago it provided a diversion from the multiple stresses of bereavement, a new job and moving home, and it remains for me a place of solitude and calm. Here I can indulge my interests and thrash out my demons. I spend a lot of time here watching birds, mammals and insects. I take photographs, and do a little survey work for the RSPB and Natural England. But I like to stop and watch the cattle too. Herds of mixed breeds of suckler cows and their calves occupy the fields from May to November, eating the lush grass and conserving the habitat for breeding waders. They have a look of contentment which I find quite comforting, so I will happily lean on a gate and watch them while they graze, chew and doze in the summer sun. And if I see a yellow wagtail fossicking around their feet then so much the better.

It's hard to think of a system of animal production that supports the needs and natural behaviour of a species of farm animal as well as the extensive grazing of cattle. In many respects the husbandry is similar to the lives of the ancestral forms of cattle. And, even better, these animals shape the landscape and, in the right circumstances, benefit biodiversity. Of course, the carbon footprint of cattle (and other livestock) and the adverse effects of the intensive systems that are more typical cannot be ignored – and we'll return to this in Chapter 11.

Not all cattle live in such an idyll, and even for the cattle on the Levels and Moors, there are various events, for example transportation and mutilation, which affect their welfare, particularly if done badly.

Sheep husbandry, on the face of it, is similar to that of suckler cows in that the vast majority are kept outside. But, as with beef cattle, there are many different systems, each with subtle differences often dictated by breed, geography and topography.

In the summer of 2021, I fulfilled a long-held ambition with a visit to St Kilda, a small archipelago lying over 65 kilometres to the west of the Outer Hebrides. It's remote and these days uninhabited. It took over nine hours to reach by fishing boat from the Isle of Harris but was worth the gruelling sea passage just to see the thousands of seabirds. But, of greater interest, even to a wildlife nut like me, are the physical remains of human occupation dating back 5,000 years. Here, around 120 people eked out a living from seabirds, from sheep, a little cultivation and a little barter until a range of circumstances and an ageing population forced the evacuation of the remaining few dozen inhabitants in 1930.

All that is left are a neat row of ruined cottages and a scatter of other stone structures (albeit with some lovingly restored by the new guardians, the National Trust for Scotland) and the sheep. Although the islanders took most of the animals when they left, some remained and they formed the basis of the thriving population present today. The resident sheep are the Soay, a 'primitive' breed (named after one of the islands of the archipelago). Small, wiry and with wool of varying shades of brown, this hardy little animal is not managed in the conventional farming sense. The sheep are part of a long term scientific study which has used the opportunity provided by isolation and minimal human intervention to investigate genetics, aging, behaviour and disease.[1]

In breaks between photographing great skuas and fulmars, I took the opportunity to watch the sheep. There were ewes with lambs in loose groups either grazing around the ruins or loafing while chewing the cud in the open pasture. Interspersed with them were rams, in groups of four to six, the typical 'bachelor herd' seen in many societies of herbivorous animal. In the late summer the ewes come into heat and mating begins. The bachelor herds break up and the rams compete for mating rights. Around the same time, as the ewes' milk production drops off and the lambs' capacity to eat and digest grass increases, the gradual process of weaning begins. All in all, not so very different to the way that most farmed sheep are husbanded. Of course, in commercial sheep farming there are interventions – the lambs are weaned earlier than they would be if they were left to their own devices, the rams are introduced to the ewes at a time to suit the farmer (and the market for the lambs), and there are routine mutilations and other interactions. But with these few exceptions, the husbandry of sheep in most systems in the UK, and hence their experience, appears to be close to what we know of the way of life of ancestral forms.

What is a grazing animal?

This chapter looks in more detail at cattle and sheep husbandry in the UK and sets out the animal welfare concerns of each. For reasons that will become apparent, I discuss beef cattle and dairy cattle separately, followed by sheep.

But, by way of introduction, it is useful to consider why grazing animals are the way they are. Grazing animals – that is, animals that eat mainly grass and other plants – are really quite different from other familiar animals. Unlike the cat (an obligate carnivore) and the dog (an omnivore), sheep and cattle have a digestive system which is capable of digesting large amounts of herbaceous material. Much plant material is low in nutrients and resistant to the digestive enzymes of animals with a short, unspecialised and simple digestive system such as dogs, cats and humans. Which is why cats and dogs don't eat much grass. In contrast, the digestive system of grazing animals is long, complex and fitted out with fermentation vats in which thousands of billions of bacteria get to work on the ingested plant material to turn it into something capable of being digested, absorbed and metabolised into energy and protein.

There are two main types of grazing animal: hind-gut and fore-gut fermenters. The rabbit and the horse are hind-gut fermenters. Their fermentation vat is an extension of the large intestine, which means that bacterial fermentation takes place after the ingested plant material has passed through the glandular stomach. This is not a terribly efficient system since the main nutrient absorption takes place further upstream in the gut (which explains why the rabbit re-ingests its night-time droppings to maximise nutrient extraction). In the other group, the fermentation vat is located before the glandular stomach – hence the descriptor 'fore-gut fermenter'. These animals are known as ruminants, and they include sheep, cattle and goats.

Everyone knows cows have four stomachs. Except that they don't. Cattle (and sheep and goats) have a single glandular stomach and a series of three large non-glandular extensions of the oesophagus (the gullet). These extensions, the largest of which is the rumen (hence the term ruminant), are receptacles for the fermentation of grass and other plant material. This is intermittently regurgitated and re-chewed to increase the surface area and hence optimise fermentation. Once fermentation is complete, boluses of material are passed into the single glandular stomach, where the otherwise undigestible precious nutrients are made available for absorption. This means that ruminants and other grazing animals can exploit grasslands. The wet and mild climate of the west and north of Britain is less suited to growing crops – the higher elevations and poorer soils, it is argued, are better given over to grass. We then use cattle and sheep to convert the grass into protein in the form of meat and milk.

As a result, cattle and sheep dominate British agriculture. Horticulture and arable farming may be much more productive and profitable per hectare, but the area given over to the grazing of cattle and sheep and to growing the crops that feed them is far greater than the area used by all other farming activities combined. Approximately 72% of the land in the United Kingdom is defined as Usable Agricultural Area. Of this area, a little over 70% is classed as common rough grazing, permanent grassland or temporary grassland. In other words, almost half the land in the UK is given over to grazing animals. This does not include the land devoted to the growing of crops such as maize, other grains and other crops to feed grazing animals. The outputs from grazing animals are economically important. Beef, dairy and sheep meat represent, respectively, 7.3%, 16% and 4.6% of the UK's gross agricultural output by value.[2]

We like our beef and dairy products. Annual consumption of beef per person is around 11.5 kg (2021), having risen steadily from a low of 9.6 kg in 1996 when the country was gripped by the BSE crisis.[3] Similarly, consumption of dairy products remains popular, with 98.5% of households buying liquid milk. Although the amount of fresh milk each of us drinks has fallen by around 50% since 1974, this is somewhat offset by an increase in the consumption of cheese and yoghurt.[4] In contrast, we are becoming increasingly less keen on sheep meat. While we consume almost three times the OECD average, the amount of sheep meat consumed annually has fallen from an average of 6.3 kg per person in 1990 to around 3.9 kg in 2021.[5]

The vast majority of cattle and sheep have a great deal more freedom of movement than most pigs and poultry. They are free to graze, wander, loaf and interact with other members of their social group. In the UK, few cattle and almost no sheep are permanently housed. And with the demise of neck tethering of dairy cattle over the winter months, even when housed, most cattle have plenty of space to exercise, feed and rest. However, there are many ways in which cattle and sheep are kept, and although I would argue that extensive grazing systems for beef cattle and sheep meet the majority of their needs and wants, none is without some welfare concerns. For instance, where sheep and cattle are out-wintered there is a risk of poor welfare if the available grazing and supplementary feed is scant and poor quality. And with the advent of very large dairy farms where zero-grazing (see below) is growing in popularity, permanent housing and lack of access to pasture constraint is becoming an issue for the dairy cow.

Increasingly, to preserve the pasture from damage, to protect cattle from the worst of the weather and disease, to ease feeding and make life better for staff looking after them, cattle of all breeds and for all purposes are in-wintered, generally in large, airy buildings. In my experience, this type of housing, while it prevents cattle from exercising and wandering for a few months each winter, also does a pretty good job in meeting their needs and probably most of their wants.

Cattle

The ancestral form of husbanded cattle is the extinct aurochs (*Bos primigenius*). Domesticated cattle have been present in Great Britain for around 5,500 years. In the absence of the ancestral form, wild or captive, it follows that the natural behaviour and social organisation of cattle is difficult to study. However, some so-called primitive breeds kept in environments with little human intervention provide us with clues.

The Chillingham herd in Northumberland, because it is generally left to its own devices, is a herd with a more or less natural sex ratio and age distribution. Some sex segregation occurs, with mature bulls living separately from cows, calves and young bulls, which form into groups of varying size,[6] suggesting that, left to themselves without human intervention, cattle would form small matriarchal groups rather like extant species of wild bovines such as the gaur of India or the African buffalo. As we will see, these social groupings are not far removed from the experience of cattle in some current extensive systems.

There are two distinct groups of cattle breeds in the UK. There are around 40 breeds of beef cattle and rather fewer breeds of dairy cattle. As you might expect, beef cattle are used to produce beef and dairy cattle to produce milk. Yet this distinction is not hard and fast. While dual-purpose breeds are essentially a thing of the past, around half of the beef we eat comes from dairy breeds. This is for two reasons – first, male calves born to dairy cows (and some of the females not suitable as dairy cows) are often reared for beef and, second, dairy cows at the end of their commercial life are also slaughtered for beef. Similarly, heifers derived from cross-bred dairy and beef breeds are often retained and reared for meat as suckler cows.

Beef cattle

The other half of British beef comes from herds of cows known as 'single-sucklers' – that is, beef cows each producing generally one calf per year which is reared from birth by the cow, weaned at about 8–11 months of age and then reared independently of its dam until it is slaughtered for beef at between 20 and 36 months of age.

Suckler herds are, in comparison to dairy herds, quite small. In England the average is around 27 cows, and there seems to be little trend over time.[7] The systems of extensive grazing that are typical of the suckler herd mimic much of the circumstances of the ancestral cattle and although the bachelor herds are absent and weaning is abrupt at a time of the farmer's choosing, their behavioural repertoire is largely unconstrained.

Other than disease, the main welfare concern for beef cattle – apart from dehorning, which is covered under dairy cattle – is castration. Although government advice is that stock-keepers should consider carefully whether castration is necessary,[8] the vast majority of male cattle not reared for breeding in the UK are castrated, and, unlike pigs, there is little evidence of any reduction over time in the proportion of male cattle that are castrated.

It is argued that castration of bull calves has significant advantages. Castration is an ancient husbandry procedure used to produce docile cattle for draught work, to reduce unwanted breeding, and to modify carcass quality. Steers are much easier and safer to handle than entire bulls. However, in much of mainland Europe the production of beef from entire bulls is the norm, although cattle rearing systems and age of slaughter are generally quite different to what we are used to in the UK.

There are three methods which can be used to castrate calves, and all of them have side effects and cause pain: a rubber ring or other device to restrict the flow of blood to the scrotum, which can only be used in the first seven days of life, by a trained and competent stock-keeper; bloodless castration by crushing the spermatic cords of calves less than two months old, with a burdizzo, by a trained and competent stock-keeper; castration by a veterinary surgeon, using an anaesthetic. Under the Veterinary Surgeons Act 1966, as amended, only a veterinary surgeon may castrate a calf which has reached the age of two months. There is evidence of pain persisting for several days after the procedure. Pharmacological methods are available to virtually eliminate the acute pain experienced by calves during the 12 hours following castration.[9] It is encouraging to see the British Veterinary Association and veterinary practices advising their clients to use analgesics above and beyond that required by the law.[10] Anecdotally, their use seems to be increasing.

Dairy cattle

The dairy cow is an important animal. Remember, fresh milk is bought by almost all households. And the UK is around 98% self-sufficient in liquid milk. But it wasn't always so important. Before tractors and cars, in terms of the economy, the horse was the dominant animal. This dominance was reflected in the training of undergraduate vets. Every aspect of the horse's anatomy, physiology and health were considered in great detail. There were weeks of lectures devoted to lameness and corrective shoeing, driven by the importance of the horse to our daily lives. But by the time I reached veterinary school in the 1970s, the emphasis had shifted – to cattle and, in particular, the dairy cow. The dairy cow was a central part of the post-war imperative to feed the nation and reduce the reliance on imported food. From the late 1940s every aspect of the development and advancement of the dairy industry was supported and encouraged – productivity, expansion, disease control, marketing and advertising. Full-cream milk was full of protein, energy and, especially in the summer, vitamins. Up until the late 1970s, primary school children were given a third of a pint of full-cream milk (a little under a fifth of a litre) at every morning break. Cheese and butter were the superfoods of the 1960s. Clever and ubiquitous advertising encouraged greater and greater consumption. It seemed that the future health of the nation rested on everyone glugging down at least one pint a day.

In support of this objective, we nascent vets were schooled in managing everything that might hinder the production of milk – the eradication of brucellosis, the threat of foot-and-mouth disease, the impact of uncontrolled mastitis, how to deal with lame cows, the latest in feeding and nutrition, and in reproduction and obstetrics. That is, how to ensure that each cow produced a large amount of milk and healthy calves.

In response to this support and encouragement, after the Second World War, the national dairy herd grew and grew. Initially herd sizes remained modest in today's terms – the average size of dairy herd in England in 1960 was 20 cows.[11] With government support and the advent of mechanisation, herd size steadily increased. It is still increasing. By 1998 it was 78 cows, and by 2020 it had doubled again to 155.[12] Now, the UK has the eleventh-largest national dairy herd, and while the total number of UK dairy cows fell from 2.6 million (1996) to 1.9 million (2018), production in 2018 was the highest annual figure since 1990 at 15 billion litres of milk.

This means that fewer cows are producing more milk per head. And on fewer farms: the number of registered dairy producers fell from 35,741 in 1995 to 12,209 in 2019, a 66% reduction.[13] Sophisticated housing and mechanisation in the form of milking machines, mechanised feeding and slurry management allow for an economy of scale. This supports keeping more cows per farm, which are looked after by a smaller and smaller number of people. And the cows, now fewer in number, are producing greater and greater amounts of milk.

The statistics bear this out. Fed on a high-energy and high-protein diet, a dairy cow produces an average of 7,800 litres a year (2019). The dairy cow is therefore very different from its ancestors, which, when suckling, would produce around 1,000 litres to rear a calf. Selective breeding, better health care and sophisticated diets have helped produce an efficient, modern dairy cow. Efficient, that is, in converting grass to milk. But it comes at a price. The former reliance on smaller, lower-yielding breeds of cow such as Ayrshire, Dairy Shorthorn, Guernsey and Jersey has reduced, with the larger, high-yielding and more demanding Holstein-Friesian cow now the dominant breed.

Like all animals selected for attributes at the extreme (high milk yield, fast growth, extremes of size, for instance), these animals require high standards of care. Sophisticated diets must be available year-round to fuel high milk yields, and unless diets and other aspects of husbandry are optimal then problems which affect the welfare of the cow can develop. These problems are seen in the high-yielding cow much more commonly than in the low-yielding cow.

Health and welfare of the adult cow

Healthy, well-fed and well-cared-for modern cows will produce large amounts of milk. Yet the physical and physiological impact of producing so much milk and the necessarily huge dietary intake can take its toll, and this can manifest in poor health and wellbeing. Many conditions beset the dairy cow, but there are three main groups which cause the most problems: mastitis, infertility and lameness. These are not discrete diseases but are convenient labels for a number of conditions which have a variety of causes.

Mastitis

Mastitis is an inflammation of the mammary gland or udder. In dairy cows, mastitis is usually caused by udder infections resulting from bacteria. Untreated, a case of mastitis can be painful, can severely affect milk yield and in rare cases, kill the cow. The basic rules to reduce the risk of mastitis in dairy cows are well understood and well known. It was drummed into me as an undergraduate in the 1970s: the cows need a dry and comfortable living area, and you must use good milking procedures, well-maintained milking equipment, an antibacterial teat-dip after each milking, and appropriate veterinary care to prevent and treat cases. These are scientifically proven to reduce mastitis to a low level. Their widespread adoption has reduced the numbers of cows affected by mastitis by a substantial amount. But mastitis is still common and a major welfare problem and results in a high level of premature culling.[14] Why is this? Achieving sustained reduction in mastitis prevalence requires multiple, detailed and time-consuming interventions both at the level of the herd and for the individual cow.[15] It is difficult to envisage that being practicable in the larger herds where staff numbers are kept as low as economically practicable.

Infertility

Profitable dairy farming relies on getting the cow in calf every year or thereabouts, so chronic infertility represents a substantial hindrance. Infertility is not, in itself, a welfare problem, but it can be an indirect indicator of poor welfare.

It might be trite to claim that healthy, contented and well-fed animals breed regularly and without any great fuss. But compare the fertility rates of the suckler cow (around 85–90% of cows will produce a calf every year) with the that of the modern dairy cow. In the UK, the conception rate has declined about 1% every three years and is now at around 40%; it is lower today than it was 40 years ago.[16] There are many reasons for this difference, but it is probably best summarised by the government's own advisory body, the Farm Animal Welfare Council (FAWC), which, in 2009, stated:

> The cow with high genetic merit for milk production produces more milk partly because of a greater propensity for losing body condition to

support milk production. This leads to a greater negative energy balance in early lactation, with more rapid loss and a slower recovery of body condition that, in turn, affects her ability to conceive. The immune resistance of high yielding cows in negative energy balance during early lactation is weak, raising susceptibility to some diseases. Cows of high genetic merit for milk production need a high level of management to ensure good nutrition, avoid extremes of body tissue loss and hence be fertile.[17]

In other words, high-yielding dairy cows have poor fertility because they are too busy producing milk to exhibit oestrus or to conceive. There appears to be a genetic component – that is, selecting for high yields can affect cow fertility.

Lameness

Lameness is a major welfare concern and one of the primary reasons for the premature culling of dairy cows, typically accounting for about 10% of culls. It causes considerable pain and distress to the cow, increases veterinary costs, takes much staff time, reduces milk yield,[18] and can also impair fertility.

The reasons why lameness develops in dairy cows are complex, which explains, in part, why lameness is such an intractable problem for the dairy industry.[19] There are many forms of lameness in the dairy cow – some are infectious, some have a substantial physiological element, and others have a genetic component. Lameness can be difficult to prevent on a modern dairy farm and can be the cause of substantial economic loss, because lame cows produce less milk. Poor-quality floors, ineffective foot trimming, poor nutrition and prolonged standing on concrete floors may precipitate the development of lameness. However, that is not the full story: higher-yielding cows are at increased risk of lameness (and other production diseases), and high standards of management are essential to optimise their health and productivity.[20] The best and the worst of dairy farms have a huge difference in the prevalence of lameness.[21] Reducing the prevalence of lameness requires hard work, investment in infrastructure and the cows' environment, routine foot trimming and good stockmanship.

*   *   *

These three groups of conditions are cited as the main reasons for premature culling of cows from dairy herds. This means that dairy cows have relatively short lives before they are removed from the herd, generally for slaughter for beef. On average, a dairy cow is removed for reasons of health, infertility or other problems at five years of age, whereas a beef suckler cow is likely to be

8–10 years old before removal. The average shorter life of the modern dairy cow is a consequence of physical and physiological stress brought about by its high productivity.[22] One could use longevity as, admittedly, a fairly crude proxy for the impact of high productivity. I use the term 'crude' because a herd with high average longevity could be masking underlying health and welfare problems that have not been addressed. It is therefore an oversimplification to state that shorter longevity is a consequence of the animal being 'worn out', but it is no exaggeration to say that a dairy cow has a harder time than a beef cow.

Of course, there are lower-impact systems which use lower-yielding, smaller cows. Organic systems claim to offer a more sensitive approach. However, the fact remains that much modern dairy farming is intensive, and increasingly so. Given the numerous threats to its wellbeing, it is hard not to conclude that the experience of the dairy cow could be considerably better if it were not for the overriding imperative to maximise the yield of each and every cow.

Mutilations

Unlike pigs, other than the various methods of identification, there are few routine mutilations of dairy cows (though most male progeny, apart from those destined to be reared as stock bulls in a bull beef system, are castrated). But since the advent of loose housing, where the cows are no longer tethered at the neck over the winter months, it has become routine to either remove the horns of dairy cows or to destroy the growing horn buds of young calves. This is to stop cows injuring each other with their horns, something which is much more likely to happen when they are housed. Although adult cattle can be dehorned, it is much more common (and easier) to remove the growing horn bud when the calf is 2–4 weeks old. The horns of adult cattle can be removed with shears, wire or a saw. Disbudding is generally done with a hot iron designed for the purpose.

Disbudding and dehorning were jobs I loathed. Almost invariably it was dirty, hot and sweaty. Either I or the farmer would get trampled in the inevitable melee (in some respects grappling calves is harder than grappling adult cattle). There would be blood spilt, both human and bovine, and everyone would end the job bruised and in a bad mood. And as for the cattle? I think they fared even worse.

I am not alone in finding this job one of the most distressing tasks I have ever had to do. The late Poet Laureate Ted Hughes wrote a poem simply entitled 'Dehorning', which provides graphic detail and does not spare the reader from the gore, pain and fear of the process.[23]

Removing horns is not like trimming a toenail. The horn, while covered in layers of insensitive keratin, is a sensitive organ served with its own blood and nerve supply. In mature cattle with large horns the base of the horn is hollow, and this space is contiguous with the frontal sinuses of the skull, which in turn are connected to the upper respiratory tract through which the animal breathes. I have often seen and heard the breath of the animal which puffs through the hole created in the sinus once the horn has been shorn off.

Analgesia, generally in the form of injected local anaesthetic, is a legal requirement in Britain. Used properly this is effective pain relief at the time of the procedure and for a few hours afterwards. But there is evidence that the pain following the procedure is prolonged. Following dehorning, indicators of acute pain such as raised plasma cortisol concentration, increased heart rate and lack of rumination are consistently recorded. These are all but eliminated where both local anaesthesia and postoperative analgesics are used. However, because these analgesics are relatively short acting, studies show that pain indicators such as reduced rumination and poor growth rates, sometimes lasting several weeks, are associated with dehorning even when combined with administration of some types of postoperative analgesic.[24] You can see it in the calves. They are listless, often immobile, and they won't eat for hours after the procedure. It is encouraging to hear calls from the veterinary profession for the use of routine post-procedure analgesics.[25]

Dehorning and disbudding are horrible and barbaric practices that need to be phased out. Of course, it is argued that housed, horned cattle will inflict much more painful injuries than would ever be inflicted during horn removal. That might be so, but it ignores an inconvenient truth: there are many breeds of cattle that are naturally polled, that is, they don't grow horns. You would think that farmers would be falling over themselves to use these breeds. But there are no wholly polled dairy breeds. Something in the genetic code of cattle means that the genes for milk yield and the presence of horns are closely linked, so any attempts to introduce polled genes invariably results in a substantial drop in milk yield. Although there is some progress in increasing the proportion of polled dairy cattle, it is slow.[26] Perhaps gene editing technology can be used to insert polled genes into the bovine genome without loss of milk yield.[27] As someone who has seen the misery that dehorning causes to cattle, and who carries the physical and mental scars of having had to dehorn cattle, this strikes me as something that is well worth pursuing. Let's hope that in a few years' time we see the end of horned dairy (and beef) cattle and an end to the need for this horrible, painful procedure.

Health and welfare of the dairy calf

In the wild, a female bovine reaches sexual maturity at around 2–3 years of age and, following successful mating, produces a calf some 9 months later.

Natural weaning occurs between 8 and 11 months of age. In contrast, on most modern dairy farms, it is routine practice to remove the calf from the cow generally between 24 and 48 hours after birth. The calves are then reared on an artificial milk-replacer until they are weaned at around 12 weeks of age. Frequently the calves are reared individually and separately in pens to reduce the risk of disease transmission, and hence they have little or no social contact. Critics say that this is inhumane, and advocate keeping the calf with the dam.

The stability of the maternal bond, particularly in mammals, is considered to be an important element of the rearing of young animals. This enforced early separation of the calf from the cow is controversial and the subject of much debate. It is not a routine feature of any other system of farming. Is it humane?

There are a number of arguments put forward in support of early calf separation. First, early separation is thought to increase profits because it allows the harvest of milk that the calf would otherwise drink. Second, separation and the subsequent feeding of calves allows the farmer to control the quantity and quality of colostrum consumed, ensuring passive transfer of immunity and hence better calf health. Third, milk release, facilitating machine milking, is thought to be facilitated by separating the calf. Finally, assuming that the mother–infant bond develops slowly in the hours and days after calving, early separation is thought to minimise the distress response for both the cow and calf.[28]

However, while suckling reduces saleable milk yield when measured over the long term, calf growth is often improved by suckling. Prolonged contact may provide longer-term benefits for calf growth and behavioural development. Further, there appear to be no health benefits from early separation for either the calf or the cow.[29] There are long-term benefits of prolonged contact in terms of sociality, fearfulness and future maternal behaviour.[30]

In summary, while initial separation appears to cause the calf some distress, the effect on the cow is transient. However, in the longer term keeping the calf with the cow for a prolonged period is clearly beneficial for the calf in terms of health, future behaviour and socialisation. It also appears to reduce the risk of the cow developing mastitis. Farmers recognising these benefits are now running dairy farms where the calf is left with the cow for up to five months. But, as you would expect, the calf's appetite for its mother's milk reduces the amount available for the farmer. One farmer estimated this loss to be around £500 per cow per year. This works its way through to the farm gate, where the retail price of the milk is at least twice that of conventionally produced milk.[31] Similar premium prices can be expected for cheese. Inevitably these are niche products, and as with the pork, poultry and eggs reared to standards designed to provide better animal welfare, discussed in

Chapter 5, the long-term success of these ventures will be dictated by the premium the products can attract rather than through gaining a substantial share of the market.

Maintaining the milk yield of a cow involves getting it to produce a calf once a year or thereabouts. One half of all calves born to dairy cows are male. If you are breeding purebred dairy cattle, the males aren't a great deal of use. They can be reared for beef but, after decades of selection for milk yield, the conformation and growth rate of male dairy calves isn't commercially attractive. That is why a substantial proportion of dairy cows are crossed with a beef bull, resulting in both male and female calves that will produce a desirable beef carcass.

The market for these surplus male dairy calves is subject to significant fluctuation. When the price is low many farmers opt to kill them at birth. Or they are exported alive for rearing elsewhere. Neither outlet helps the reputation of dairy farming, particularly as those exported to the EU are often destined for veal systems (though note the government's intention to prohibit exports of live animals other than those destined for breeding).[32] There have been various initiatives, some farmer-led, to reduce the surplus of male calves. These have had mixed results, simply because, depending on the market, it is often cheaper to have the calf killed shortly after birth than to rear it.[33] However, the development of technology to produce bovine semen differentiated by sex with a high degree of accuracy gets around a great deal of this problem. Now that the technology is proven, increasingly dairy farmers are selecting semen for artificial insemination which will produce a female calf of a dairy breed, thereby reducing the numbers of surplus, low-value, purebred male calves.[34]

Do increasingly large dairy herds threaten animal welfare?

It stands to reason that as the number of cattle (or other animals) that one stockperson is responsible for increases, so the care that can be lavished on each individual will decrease. Along with public opposition to having large numbers of farmed animals close to where they live, this is the reason why a number of animal welfare organisations campaign against so-called megadairies. These large herds often use a system of feeding known as zero-grazing, where the forage is cut and carried to the cows, allowing year-round housing. Compassion in World Farming believes that 'cows belong in fields' and campaigns for more traditional husbandry, particularly organic systems where it is argued that the care of the animals is front and centre of the operation.[35]

The productivity and the welfare of the high-yielding dairy cow relies on providing high-quality care and husbandry. It is difficult to see how this can be consistently provided when dairy staff have sole charge of 250 cows and upwards. Of course, mechanisation, sophisticated housing and high-quality diets help, but expert government advisers consistently emphasise the value of good stockpersons in ensuring animal welfare: 'The stockman has a unique role within livestock farming in ensuring high standards of animal welfare.'[36]

Despite misgivings about large dairy herds, there is evidence that the absolute number of cattle per herd is less important than the nature of housing conditions and management practices.[37] It can be argued that being part of a very large herd actually has the potential to benefit an individual cow's welfare when a team of experts, including a nutritionist and a specialist veterinarian, can be employed to monitor health, disease prevention and control.

This issue is unresolved and is likely to form one part of the argument deployed against the building of mega-dairies, along with arguments about nuisance and threats to the environment. One aspect that might hold sway is evidence that the physical wellbeing of the dairy cows is better when they have access to pasture, and that dairy cows appear to want access to pasture. A wide-ranging review concluded that cows on pasture-based systems have lower levels of lameness, hoof pathologies, hock lesions, mastitis, uterine disease and mortality compared with cows on continuously housed systems. Pasture access also had benefits for dairy cow behaviour, in terms of grazing, improved lying/resting times and lower levels of aggression.[38] Interestingly, when given the choice between pasture and indoor housing, cows showed an overall preference for pasture, particularly at night. The reviewers concluded that there are animal welfare benefits from incorporating pasture access into dairy production systems.

One further aspect which pertains to large herds, continuously housed or not, is the social nature of the species. Cattle, like most grazing animals, are social creatures. Given the opportunity, they will form small matriarchal groups of up to 20 animals of mixed ages, although young bulls will often form separate groups of their own. There is growing evidence that cows will form strong bonds with each other to the extent that when that 'partnership' is broken both individuals will exhibit signs of distress and reduced milk yield.[39] In small dairy herds these partnerships are often relatively easily maintained, by design or happenstance. However, the recent increase in intensification of the UK dairy industry has led to increasing numbers of dairy cows being housed in large, dynamic groups. The preservation of these partnerships (assuming they have been noticed) becomes more difficult as cows are moved into different groups for management purposes. For instance, some cows may move groups two or three times during a single year.

Sheep

Across Britain, sheep are kept on a variety of types of agricultural land – from the lush and comfortable grazing of the lowlands to lands on the very edge of the country where they eke out a living on rough upland grazing. Over centuries, farmers have bred sheep for the physical and behavioural characteristics that ensure the breed is well adapted to the particular environment in which they are kept. This is important in the uplands, where hardy breeds must withstand tough winter conditions. In the lowlands, where the living is easier, a greater emphasis has been given to traits such as fecundity, wool quality and carcass conformation. Whatever the breed and whatever the location, sheep are reared for two main purposes – for meat and wool.

Britain has the second-largest national sheep flock in Europe, behind only Spain. The number of breeding ewes is just under 14 million (2020), although this has fallen from a high of over 20 million (1998).[40] Historically, sheep numbers have been supported by generous headage payments, particularly in the uplands, but the decoupling of payments from production and the adoption of a system of area payments has reduced the incentive to keep large numbers. This, combined with the British eating less and less lamb and mutton, and a fall in the demand for wool, has squeezed margins. Exports of lamb particularly to France and Italy have helped to maintain the industry somewhat, but this has not been enough to reverse the decline, and the trend in sheep numbers remains downwards. Nevertheless, sheep production remains important, particularly as a source of income and employment in the uplands.

Mutilations

Other than concerns about disease, transport and slaughter, there are two main welfare issues important in sheep husbandry. The first is mutilations, of which two are carried out routinely: castration and tail docking. Many male sheep are castrated, and it is common practice to dock the tails of many breeds of sheep. Obviously, sheep intended for breeding are not castrated, and hill breeds such as the Scottish Blackface and Welsh Mountain are rarely tail-docked.

The testicles and the tail of the sheep are not part of the anatomy by accident: they evolved for a purpose. It can be argued that, like other mutilations, the castration and tail docking of sheep are, ultimately, performed for the convenience of the shepherd. However, both have a practical purpose. Hence, as with other mutilations, it is legitimate to ask two questions:

- Does it need to be done?
- Can it be done better?

Castration has been performed in sheep since domestication. While it is governed by detailed rules, any of three main methods (rubber ring application, bloodless castration using a burdizzo device, or surgical castration) may be performed by lay people and without anaesthetic. Restrictions apply to each method after a certain age.[41] There is evidence that all methods cause pain.[42]

Castration is justified with the following reasons:

1 Castrated males are less aggressive and easier to manage.
2 When all males are castrated there is no risk of early-maturing ram lambs mating with ewe lambs or their mothers.
3 It avoids ram taint (unpleasant tastes caused by testosterone and skatole in the fat).

Docking appears to be a later practice, not appearing in records until the sixteenth or seventeenth century. Its main purpose seems to have been the reduction of faecal or urine contamination in the longer-woolled breeds that began to be prevalent at that time,[43] which increases the risk of blowfly strike. Interestingly, the more 'primitive' breeds tend to have shorter tails and consequently the risk of contamination and fly strike is reduced. There are a number of methods of tail docking including the use of rubber rings to restrict blood supply, hot cautery and the 'cold' knife. As with castration, there is detailed legislation which governs the practice.[44]

The application of the 3Rs approach to castration and tail docking offers alternatives. The principles of the 3Rs (replacement, reduction and refinement – explained in detail in Chapter 10) were developed over 50 years ago, providing a framework for performing more humane animal research. However, in practice the 3Rs can be applied to any procedure involving animals in all circumstances. For example, the need for castration can be reduced by managing lambs in single-sexed groups, and the procedure can be refined by the use of analgesics – which reduces plasma cortisol, a substance produced in response to noxious stimuli. Similarly, the use of analgesics with tail docking reduces the pain response. Selective breeding for shorter tails offers the prospect of a reduction in the need for tail docking.[45]

The BVA now recommends that castration of ram lambs should only be carried out when there are no suitable alternatives available, and that to reduce the need for castration, farmers and vets should work closely and tailor their farm animal health and welfare plans.[46]

Lameness

The second welfare concern is lameness. It is possibly the most serious welfare problem in sheep and a source of serious economic loss.[47] It has been estimated that at any one time around 3 million sheep in the UK are lame.[48] Lameness in sheep has a variety of causes, the leading one being one or other type of bacterial infection. There is no doubt that lameness in sheep is painful and debilitating.

As with dairy cattle, the treatment and prevention of lameness in sheep requires close monitoring, hard work and attention to detail. Infectious forms can be eliminated from flocks but, in conjunction with other measures such as vaccination, prompt treatment and the culling of badly affected individuals, this requires careful attention to biosecurity, particularly when new sheep are added to the flock. Despite efforts up to now, the prevalence has remained unacceptably high. This prompted the government's own advisory body to recommend that government should work with industry to develop a national strategy, that surveillance should be done to determine trends in lameness over time, and that a target to reduce the prevalence of lameness in flocks should be introduced.[49] There has been some progress with farmers adopting new regimes involving prompt treatment of lame sheep and vaccination rather than the traditional approach of routine flock foot trimming,[50] but reducing the prevalence down to the 2% target requires further efforts.

Summary

Sheep and cattle, particularly when grazed extensively, appear to have their physical needs met and have the freedom to express many of their normal behaviours. Even when they are housed, given sufficient space and a well-maintained environment, it is hard in most cases to be critical. However, as we have seen, there are a number of areas where welfare may be compromised. In particular, the necessity for the routine mutilation of cattle and sheep, and the way in which the procedures are carried out, needs to be challenged, particularly when there are alternatives and means of reducing the attendant pain. The application of an approach to mutilations in sheep similar to the 3Rs is an encouraging development, and the prospect of more polled cattle may mean an end to the frankly barbaric practice of dehorning. Similarly, the running sore of sheep lameness can be addressed by adopting new practices proven to work.

Chapter 3 raised concerns about transport and slaughter, and there is no doubt both could be improved with a bit of thought and commitment simply by using best practice and existing technology. Reducing distances to slaughter would be a good start.

The welfare of the dairy cow is a different matter. Despite being of individually high value, and despite the economic importance of the dairy sector, there are a number of concerns which, while they are not being ignored, suggest that welfare is often compromised. This applies to the high-yielding and intensively managed herds and has the potential to undermine the reputation and acceptability of dairy products. Premature culling of adult cows in response to lameness, infertility and mastitis is closely associated with the selection and husbandry for high milk yield. Dairy farming with lower-yielding, less physiologically stressed cows, and with access to pasture, seems more likely to meet the needs and wants of the cow.

The evidence about the very large dairy herd being detrimental to animal welfare is equivocal, but the impact on social structure and bonds where groupings are dynamic suggests that large herds may affect wellbeing. Finally, there is evidence that the permanent housing and zero-grazing of dairy cows can lead to poor welfare. That cows prefer the opportunity to graze is surely evidence enough.

Where does this leave sheep and cattle in relation to the Five Domains? Let's remind ourselves. The domains are nutrition, health, behaviour, environment and, finally, the animal's mental state: it is an integration of the first four that dictates the fifth.[51] Despite that there is no one single measure, using the Five Domains model, the systems of husbandry described in this chapter are more likely to be judged favourably than many of the more intensive systems discussed in Chapter 5. It should be noted that this excludes the high-yielding dairy cow.

It is hopefully becoming obvious that, unlike some other systems of livestock farming, I am a fan of extensive grazing, particularly where the attendant welfare issues are addressed and there are demonstrable benefits for biodiversity and bio-abundance. And, to be clear, only certain systems of grazing aid conservation. I am conscious, however, that advocating extensive grazing does not sit well with concerns about our changing climate. This is not a book about global heating, but Chapter 11 will consider whether certain types of extensive grazing, with their associated biodiversity benefits, can be compatible with measures necessary to limit climate change.

5

Pigs, Poultry and the Rest

Pigs *and* poultry? Grouping together grazing animals like cattle and sheep, as in the last chapter, makes sense. But pigs and poultry? Yes, it is blindingly obvious that they are different and, no, I won't insult you by listing the differences. In farming, however, there are good reasons why they tend to be considered together.

First, although there are lots of smallholdings with a handful of chickens and rather fewer with pigs, these animals tend to be farmed intensively and in large numbers.

Second, although both are omnivores, the commercial diets of pigs and poultry are both primarily derived from grain – they are not grazing animals and do not have the capacity to digest large amounts of fibrous plant material. The digestive system of the pig, at least functionally, is more similar to that of the chicken than it is to cattle or sheep. Agriculturalists tend to describe pigs and poultry as the monogastric species (that is, having only one stomach), in contrast to ruminants with their multi-chambered digestion vats.

Third, the vast majority of commercially reared poultry and a high proportion of commercially reared pigs are housed for part of or all of their lives.

For completeness, at the end of this chapter I have included short sections on other intensively reared farmed birds – turkeys and ducks – and also on farmed fish.

Pigs

- 'Pigs are fascinating, intelligent and curious creatures.'
- 'Given the opportunity pigs are scrupulously clean.'
- 'Pigs are filthy animals.'

Everyone has an opinion about pigs. And often these are poles apart. While derided as dirty animals by many, others attest to their similarity to humans and their cleanliness. Human and porcine similarities are so great that the pig is the favoured model for research into skin healing. Pigs are the first and perhaps may be the only practical candidate for xenotransplantation of solid organs such as hearts and kidneys.[1]

My favourite pig quote is one attributed to Winston Churchill: 'Always remember, a cat looks down on man, a dog looks up to man, but a pig will look man right in the eye and see his equal.'[2] There is no doubt pigs are intelligent. For instance, studies have shown that pigs are capable of being trained via operant conditioning (see glossary) to obtain light and additional heat for their enclosure, and to acquire feed. Pigs can discriminate between the size and shape of objects and can respond discriminatively to human verbal and gestural commands. Recently, pigs have been trained to manipulate a joystick that controls the movement of a cursor displayed on a computer monitor.[3]

Knowledge of the pig's intelligence hasn't diminished our taste for its meat. Although pork was, for many years, the world's favourite meat, steady growth in chicken meat production has meant that it has lost that position. But global production of pig meat is around 112 million tonnes (2021), and there are projected increases in production until at least 2029.[4]

Britain is a relatively small producer, and despite our liking for pig meat we are only around 53% self-sufficient, with the vast majority of the shortfall being made up by imports from the EU. We each consume on average 16 kg of pig meat per year (2021), and this has fallen slightly from a high of 18 kg in 2007. In comparison, our average consumption of chicken meat has risen from around 22 kg per person per year in 2008 to 27 kg in 2021.[5]

As we saw in Chapter 3, pigs are social creatures and, given the opportunity, will form extended family groups and forage and root in the ground for a substantial part of each day. That is not so different from the behaviour of the ancestral form of the domestic pig, the wild boar, and close to what might be observed in the traditional low-input, low-output pig husbandry systems that were once practised the world over. The free-roaming village pig has almost gone from Europe, but they are still a feature of subsistence farming in Southeast Asia, Africa and Central and South America. These animals live in and around villages in small groups, foraging for scraps with minimal management and intervention.

In the early 1980s, I got a job in Belize. I was one of four veterinarians in the country – all working for the government. We were trying to build an infrastructure for animal health while providing a service to farmers. Newcastle disease (a respiratory disease of poultry), rabies and classical swine fever outbreaks were common and, despite our best efforts, none showed much sign of being under proper control. In 1982, I was sent up to the north to deal with an outbreak of swine fever in village pigs, and I spent several months running a vaccination programme.

It was completely different to what I was used to in Britain, where, when I was vaccinating cattle against brucellosis, the farmer would have the animals ready and penned. Imagine a rodeo featuring only pigs. To start vaccinating, my team and I would pitch up in a village and fan out looking for pigs. If we were lucky and the message from the extension staff had got through, the villagers would have penned the pigs, and within the first hour or so we would have injected three-quarters of the animals. But then we would spend the next hour trying to catch the last 15 or so. Sometimes, the message didn't get through and we would be faced with 100 or more pigs on the loose. Which is when the rodeo would start.

The sound of a squealing pig carries, and the ones we hadn't yet caught soon got wise to what was going on – they would take to the bush or hide at the other end of the village. Then all of a sudden it seemed that every able-bodied male was chasing a pig. Other times, the whole village appeared to have turned just to watch me, the gringo, chase a pig in an out of the bush, in and out of houses, until eventually it was cornered, caught and vaccinated. We didn't catch them all – we once cornered a pig in an outhouse only for it to put its head down, charge, break through the wooden wall and head for the horizon, not to be seen again that day.

It was great sport, but it wasn't much of an advert for efficient agriculture. Although they could run like a greyhound, twist and turn like a ferret and squeal like a steam whistle, the pigs were small, often malnourished and slow growing. Village pigs live in close proximity to the latrines, thereby ensuring the effective but undesirable transmission of zoonotic parasites. As well as dealing with disease control, I was also responsible for meat hygiene and inspection. On several occasions, the inspectors and I found cysts of the tapeworm parasite *Taenia solium* in village pig carcasses presented for human consumption. This is a zoonotic parasite, similar to the beef tapeworm, except it cycles between pigs and humans. Infection is inapparent in the pig but, unlike the beef tapeworm, it can cause severe illness in humans.

Managing feeding, breeding and disease in village pigs is almost impossible and in most developed countries it is a thing of the past. With the exception of small numbers of pigs given access to oak woodlands to forage on acorns ('pannage'), the genuinely free-roaming pig in Britain is no more. Although smallholders often keep a few breeding sows, these are kept behind wire fences or are housed.

Modern commercial and more intensive systems are quite different. Sows are expected to produce, on average, a little over two litters per year, and the most productive units will produce 26 weaned piglets per sow per year. In contrast, the wild boar generally has one litter per year averaging between four and six piglets each time.[6] The village pig would be similarly unproductive.

The huge increase in productivity of the domestic pig in comparison to the wild boar and the village pig is a product of a number of factors: genetic selection for litter size and unseasonal breeding, improved nutrition, effective disease control, bespoke housing and a near elimination of predation. This, combined with sophisticated selection for desirable carcass conformation and a fast growth rate, means that each sow can be relied on to produce a great deal more meat from its progeny than ever could be expected from the ancestral form. This undoubted achievement means an abundance of relatively cheap pig meat. But has the pig paid too heavy a price? Perhaps. Apart from obvious things like disease, there are a number of significant welfare concerns.

Mutilations

There are three main mutilations of growing pigs and one inflicted on outdoor sows. First, castration (the surgical removal of the testicles) is a routine procedure primarily to reduce the risk of boar taint developing in the maturing male pig prior to slaughter. Pigs in Britain are now rarely castrated, because the genetics of the modern pig means the vast majority reach the desired slaughter weight before puberty and before levels of the associated sex hormones risk tainting the meat. Where castration is practised, there are rules about the age at which it can be done and the use of anaesthesia and analgesics after seven days of age.[7] This substantial reduction in what was once a routine mutilation sits in contrast to other mutilations of pigs and the routine castration of lambs and calves discussed in Chapter 4.

Second, the removal of the 'needle teeth' (the deciduous or temporary canine teeth and outer incisor teeth). Piglets are born with eight fully erupted 'needle teeth', which may be considered as weapons for neonatal competition among siblings. There is an adaptive benefit in being born with these teeth. Getting secure access to the sow's teats is competitive, and with large litters there aren't enough teats to go round. But as well as being useful to ward off your siblings, these little teeth can damage the sow's sensitive skin around her teats, making her less likely to suckle. To limit the risk of injury it has been common practice to clip these teeth down to almost gum level in the first few days of life. More recently, electrically powered grinders have been adopted in the belief that they cause less trauma. However, both clipping and grinding expose the sensitive core of the tooth. There is evidence that in the short term both interventions cause changes to the piglets' behaviour suggestive of severe pain.[8] In the longer term both cause lesions such as pulp cavity opening, fracture, haemorrhage or abscess.[9] However, the changes to piglet behaviour are more marked and the lesions are more severe after clipping than grinding, suggesting that the latter is less painful and causes less tissue damage. Routine clipping or grinding of teeth is not permitted in Britain and, according to the government's *Code of Practice for the Welfare of Pigs*, 'if tooth reduction has to be performed it should always be considered a last resort'.[10] There appear to be no data on the proportion of piglets that are tooth-clipped, but, despite the law, anecdotal evidence points to it being a very high number.

The third mutilation of growing pigs is tail docking. Piglets' tails are docked to prevent the serious welfare problem of tail biting, which, when an outbreak occurs, can cause severe injuries throughout the affected group. Where tail biting is not controlled, the injuries can be so severe that pigs may need to be euthanised. The tail of the piglet may be docked by a competent person at up to seven days of age (although a veterinarian may dock older animals). To reduce the risk of tail biting, docking was a routine on most pig farms for decades. However, the law changed in 1994 and routine tail docking is now no longer permitted.[11] Government guidance now states that 'tail docking should only be used as a last resort, after improvements to the pigs' environment and management have proved ineffectual in preventing tail biting'.[12] Written approval from a veterinarian is a legal requirement before any tail docking can be done. Approval should only be given once a risk assessment demonstrates that tail docking is necessary and unavoidable for the particular batch of pigs in question or for a fixed period of time. More regular reviews are recommended if tail docking has been used on a farm for a sustained period of time, and there is an expectation that management procedures or changes to the system will be required.

These regulations reflect the evidence that tail docking is associated with signs of acute pain and stress in piglets and adverse consequences throughout the period up to weaning, including on their relationship with humans.[13] The absence of a tail obviously reduces the risk of tail biting and helps reduce the risk of fighting. However, campaigners argue that it is the barren environment in which many pigs are kept and the associated lack of sensory stimulation which precipitates aggression between undocked pigs. Other factors known to increase the risk of aggression between growing pigs include pen structure and cleanliness, the quality of the diet, thermal comfort, air quality and light, health, fitness, and competition. Ensuring that each of these factors is optimal requires good facilities and good husbandry.

There is no doubt that until recently the law was routinely ignored, with the proportion of pigs being tail-docked in the UK averaging 82% in 1999. However, it is clear that the British pig sector is making an effort. Assessments of nearly 3 million pigs over a period of three years showed a reduction to 70% in 2013–2016.[14] It's not rapid progress, but it's a start. It is worthwhile noting that in several European countries tail docking is rarely practised, with some countries recording no routine tail docking at all – demonstrating that rearing undocked pigs can be done provided there is the will and suitable investment in skills and facilities.[15]

The final mutilation is nose ringing. One or more metal rings are attached to the sensitive part of the nose of sows kept outdoors to try to reduce rooting of the ground. The pig has an additional bone in the nose known as the *os rostri* which allows it to root for long periods without injury. A nose ring inhibits this behaviour. As well as pain when the ring is inserted and the stress of handling, nose rings work by causing discomfort and pain as the sow goes to root with her snout, causing her to stop. Studies have found that ringed sows spend more time standing but otherwise inactive and are observed straw-chewing, vacuum-chewing and digging for longer periods than un-ringed sows, suggesting that nose ringing affects welfare.[16]

The pig is perhaps the most mutilated of all farmed animals. While there are clearly efforts to reduce the amount of mutilation, progress is slow. It is dispiriting to note that although two of the routine mutilations are effectively illegal unless there is an overriding need, both tail docking and tooth clipping are still widely practised. It is also disappointing that so many sows, having been released from confinement and allowed to forage in the open, are prevented from rooting by having rings placed in their noses.

Confinement and barren environments

The post-war intensification and expansion of pig farming meant that sows were being kept in larger and larger groups. However, sows in large groups, particularly when housed, are inclined to fight and cause each other severe injuries, especially around the time when they are fed. In an attempt to address this problem, the sow crate was introduced. A less popular system where the sow was tethered by the neck or around the girth was introduced at around the same time.

From the 1960s an increasing proportion of breeding sows in Britain were closely confined during pregnancy. Once a sow was known to be pregnant ('in pig'), it would be put in a crate or tethered for around three months until it was ready to farrow. The crate was just long enough for the sow to lie down but not wide enough to allow her to turn around. A tether would be similarly restrictive. The sow would be fed a diet sufficient for pregnancy but generally this would be given just once or twice a day. The advantages, for the farmer, were that one could guarantee that the sow was getting the right amount of food and that there would be no fighting.

However, there were disadvantages. In comparison with loose-housed sows, sows in crates showed an increased amount of lameness, probably associated with poor overall fitness caused by confinement and lack of exercise. Skin ulcers could be common unless stockmanship was good. There was another, potentially more serious concern: many sows in crates developed stereotypical behaviours. They would chew the bars of the crates to the extent that their teeth would wear flat, or develop a weaving motion sometimes for hours at a time. This was interpreted by animal behaviour scientists as a coping mechanism developed in response to the desire to forage being thwarted, something that a free-range pig would devote hours to every day. Tethering or crating pregnant sows was eventually banned in Britain in 1999, and in the rest of the EU by the end of 2012.

The vast majority of sows in Britain are now kept in free-range outdoor systems. They enjoy much better musculoskeletal health, they are much less likely to develop stereotypies, and, unless they have a nose ring applied, they can root in the ground. But while they are not now confined during pregnancy, farrowing crates are still widely used. Shortly before she is due to give birth ('farrowing'), the sow is typically moved to a farrowing crate. This is similar to a sow crate except that there is space to the side for the piglets. Bars keep the sow out of the piglets' lying area to prevent crushing. And, just as in a sow crate, the sow will be frustrated from being unable to exhibit normal behaviour. The urge to nest-build in the few days before giving birth is strong, and close confinement and a lack of bedding material must cause her considerable frustration.

Once weaned at around six weeks of age, the piglets are removed from the sow (allowing the sow to come back into oestrus, to be inseminated and become pregnant once again). These growing pigs are then grouped together with other litters and reared until they reach slaughter weight at between four and six months of age – which for the British pork market is around 80 kg liveweight. Bacon pigs are grown for longer and reach a heavier slaughter weight. Growing pigs can be housed in yards and bedded on straw, kept in slatted or concrete-floored pens, or penned outside.

Concern for the welfare of pigs confined in barren environments such as concrete-floored pens led to a legal requirement for all pigs to have permanent access to enrichment materials which provide them with the opportunity to engage in proper investigation, manipulation and foraging activities. According to the government's welfare code:

> enrichment materials should enable pigs to fulfil their essential behavioural needs without compromising their health. They must be safe, hygienic and should ideally have the following characteristics: edible, chewable, investigable and manipulable.[17]

Simply providing a football, a length of plastic pipe or a chain suspended from the ceiling is considered insufficient, since the pigs soon become uninterested and the risk of tail biting is not reduced. Copious amounts of straw and other fibrous materials appear to work much better, as they keep the pigs occupied for far longer periods.

Should you ever decide to keep pigs, use something designed for the job. Don't use a car tyre as a toy for pigs. I speak from experience here. As a newly qualified vet I once spent two hours sawing a rubber car tyre off the midriff of a pig that had attempted to walk through the tyre once too often. The pig had got stuck, like Winnie the Pooh, half way along. Needless to say, neither I nor the pig thought this was particularly funny. That it was a steel-braced radial made it all the more difficult.

What next for pigs?

Where does this leave pigs and the modern pig farm in relation to the Five Domains? The domains, remember, are nutrition, health, behaviour, environment and mental state. It is an integration of the first four that dictates the fifth domain, the mental state of the animal.[18] While nutrition is generally of a high order and disease management is generally improving, there are sufficient concerns about the environment in which the pigs are generally kept and the mutilations that are routinely carried out to indicate that in many cases, the mental state of pigs is compromised to some degree. The free-range sow which is able to farrow and rear her piglets unconfined is likely to have good welfare (though many will still bear nose rings and may be out in inclement weather – albeit with shelter nearby), and she is certainly in better physical health than her predecessors, who spent most of their adult lives confined in one form of a crate or another.

We should not forget that many sows are confined in a farrowing crate for up to five weeks, and are thereby prevented from displaying most normal forms of behaviour associated with farrowing, such as nest building. These sows will undoubtedly have a poorer mental state. John Webster, former professor of animal welfare at Bristol University, neatly summarised the predicament of the confined sow: 'Sows on concrete in confinement stalls suffer abuse according to all the Five Freedoms.' These include freedom from hunger, discomfort, pain and distress, and the freedom to express normal behaviour. Sows that have to lie on concrete flooring can experience excessive heat loss and chronic physical discomfort, while this same concrete flooring can contribute to foot injuries, joint pain complications and skin lesions owing to prolonged contact with an unusually hard surface and chronic inactivity.[19]

The welfare of growing pigs after weaning and hence their mental state will be affected by the quality of the environment. The adoption of straw yards, along with the recognition in law that the pig has needs beyond the purely physical and that the environment provided must take account of its behavioural needs to protect its welfare, is quite a step forward.

What are the implications for you, the consumer, assuming you still want to make decisions about which meat you eat rather than stopping altogether? In summary, the lot of the commercially reared pig is not great. The combination of confinement, mutilations and a barren environment (admittedly not visited on all pigs in all systems) is sufficient to make me think twice. But, as ever, there is an alternative. There are a number of enterprising farmers producing pork from systems which actively promote their welfare credentials such as enriched environments, a lack of tail docking, no use of farrowing crates. Of course, this comes at a cost. The output from these enterprises is relatively low and the pigs take longer to reach slaughter weight. These systems represent a small proportion of total production, so expect prices to be 2–3 times greater than at a supermarket. But it will taste better, and you'll probably eat less of it.

Many local butchers buy direct from smaller pig farms, and gentle quizzing will help you get more information about how the pigs are reared. As might be expected, getting information about pork in pre-prepared meals, in restaurants, and in relation to imported meat is much more difficult but not impossible. There is no harm in asking.

Poultry

The poultry sector is large, complex, sophisticated and successful. Most production in the developed world, and increasingly throughout the rest of the world, is in the control of a small number of horizontally and vertically integrated companies controlling many aspects of production including hatcheries, breeding, feed mills, farms and slaughterhouses. However, we are less interested in the corporate structure of the organisations that grow, process and market the eggs and meat than in the way the birds are reared and cared for.

Apart from the breathtaking scale, one fact above all stands out, and that is the difference between the two arms of the industry. One is dedicated to the production of birds for eating and the other to the production of table eggs. The two types of enterprise have diverged so much that there is now little in common between them. Nowhere is this more evident than in the type of bird used. As a result of sophisticated breeding and selection over many generations since the 1950s, the types of bird have diverged into different strains, one used for meat ('broilers') and the other for laying eggs ('layers'). The performance, behaviour and morphology of both types are so different that they might as well be separate species. Except they aren't.

Every breed of chicken, from the tiny 500-gram Malaysian Serama to the Buff Orpington, weighing in at up to 6 kg, and every commercial strain, from the Hisex Brown hybrid laying hen to the Ross 308 broiler chicken, owes its ancestry to the red jungle fowl.[20] This bird, as the name implies, is a bird of the 'jungles' of Southeast Asia. It is a social species that usually aggregates into small groups with one male and several females occupying a regular home range, or territory during the breeding season. Within their home range they have regular roosting sites, where they sleep high in the trees at night and rest during the hottest part of the day.[21] Group size rarely exceeds 20 individuals.

While it could be argued that the smallholder with 15 hens and one cockerel might be quite closely replicating the social structure of the ancestral form, the same cannot be said for the vast majority of commercial laying hen and broiler units. Along with other impacts of keeping chickens in large groups, that lack of a social structure has had a profound effect on their lives and experiences.

The broiler chicken: the source of ubiquitous animal protein

In Chapter 1, we touched on the ubiquity of chicken meat and how, for many people, it is now barely thought of as being derived from an animal. And yet the number of birds produced for the table is colossal. Let's remind ourselves of the sheer scale of broiler production. In Britain, we produce close on 1 billion broilers every year. That means almost 20 million chickens are killed and consumed each and every week, and that doesn't include imported chicken meat. It equates to around 15 whole broilers per person each year. And if we exclude people who don't eat chicken meat, such as vegans, vegetarians and infants, then it's a great deal more. With production and consumption on that scale, surely we are honour-bound to find out more about how it's done.

In Britain, broilers aren't reared in cages; the vast majority are reared in large sheds with a floor covered in a deep litter, which is usually wood shavings. They are brought in their thousands from a hatchery and placed there as day-old chicks, kept under carefully controlled conditions, fed a continuously available scientifically formulated diet and, depending on the market, collected and slaughtered at anything between 30 and 40 days of age.

It hasn't always been that way. A meat bird in the 1950s took between 12 and 14 weeks to reach a suitable size. In 2022, a similarly sized bird takes about a third of that time. Sophisticated breeding and genetics, combined with a controlled environment, effective disease control and a carefully formulated diet, produces a modern broiler that is fast-growing and docile, with a large skeleton and a muscular profile.

At the start of the growing 'cycle', the day-old chicks don't take up much space. The birds are placed at a density of 18–20 birds per square metre, which means that when they are close to a typical slaughter weight (1.5–2.1 kg), the stocking density can get up to 42 kg of broiler chicken per square metre, which takes up a high proportion of the available floor space.

As well as being selected for fast growth, the broiler chicken is rather passive. Throughout its short life, it doesn't do a great deal more than stand in the shed to breathe, eat, drink and defecate. Food and water are there in front of it, all of the time, never much more than a metre away. The temperature and humidity are controlled with heaters, ventilation, and in some cases air-conditioning. The only thing the bird is expected to do is to remain alive and grow. And grow they do. The combination of the broiler's innate ability coupled with the unlimited amounts of food close at hand means that the all-important feed conversion ratio (FCR) is extremely low, second only to that seen in some species of farmed fish.

It is not by chance that the modern broiler is docile. It is the result of generations of careful selection. If you want fast growth, you don't want the birds to be curious, active or aggressive. Exploring the environment and interacting with other birds uses up energy and protein that can be laid down as muscle, bone and fat. It's better to be indolent and spend your time eating – and growing. Aside from the very different physical characteristics of the broiler and the egg-laying hen, it is this docile behaviour that sets the two strains apart. Much of the typical behavioural repertoire of the jungle fowl seems to have been lost from the broiler, while more of it has been retained in the egg-laying hen. Of course, given that most broilers are killed before they reach six weeks of age, perhaps the opportunity to express most forms of territorial and sexual behaviour never arises: the broiler is, despite its size, an infant.

Provided the environment is maintained at the optimum temperature and humidity, the building is properly ventilated, and food and water is of the right quantity and quality, then birds reach market weight in the expected time with little problem. Conditions have to be *just* right, however. A former colleague once likened the modern broiler chicken to a winning Formula One car: it's sophisticated (after generations of selective breeding), it's fast (it grows rapidly), and it wins (it is profitable). But it needs great care, attention and the very best supportive technology to perform at its peak. If the car doesn't get the attention it needs, it won't go or it will break down. Just like a poorly supported broiler.

What are the conditions that cause problems for the 'Formula One' broiler? The industry has a good record on infectious disease and parasite control, but there are a number of non-infectious conditions and problems which beset the rearing of broilers – and, for the most part, these are associated with the rapid growth of the animal and the stocking density at which they are kept.

In the late 1980s, I was a newly recruited, eager and ambitious government veterinarian based in Aberdeen. Although my job was mainly dealing with outbreaks of brucellosis and tuberculosis in cattle, we investigated welfare complaints in all species of farmed animal. After I had been in post for about two years, the offices of the then State Veterinary Service began receiving reports of broiler chickens affected with a high proportion of skin lesions. I was asked to investigate local reports. It soon became clear there was an industry-wide problem but that the prevalence of skin problems varied from farm to farm. Many farms, particularly those that were older, less well-insulated and with older equipment, struggled to provide a suitable and safe environment for the rearing of broilers. This was not normally the case in the more modern, purpose-built buildings.

The environment for the intensively reared broiler must be closely controlled and monitored. The temperature, particularly in the first week after hatching, must be maintained within close parameters. As the birds grow and develop the physiological capability and gain sufficient mass to regulate their own body temperature, the need for such close control diminishes. However, the metabolism of the growing birds produces a great deal of moisture. To avoid the build-up of condensation, particularly on the litter, controlled ventilation is required. In modern purpose-built broiler houses, condensation can be controlled – insulation and ventilation are sophisticated. In older houses, condensation can be a problem – meaning that the litter can become very wet. Similar problems can develop when old, poorly designed drinkers are used. If this is combined with concurrent disease or the feeding of a suboptimal diet it may precipitate diarrhoea. This in turn can increase water consumption and compound the problem. These factors acting additively may cause a deterioration in the litter to the extent that a form of contact dermatitis can develop. This is seen most often on the soles of the feet (pododermatitis), the tarsus or ankle ('hock burn') or the skin of the breast ('breast blister'), which are the parts of the body in closest contact with the litter. Reductions in the prevalence of the various forms of contact dermatitis can be achieved by attention to diet, house design and facilities, particularly by installing modern drinkers which are less likely to leak.[22]

In Aberdeenshire, I worked closely with the more receptive representatives of the industry, who, to their credit, were keen on finding solutions. I visited several dozen broiler farms, making repeated visits to monitor progress through the growing period. Some farms had significant problems, some had almost none. In most cases, the reasons were obvious – poor management, poorly constructed and old buildings, concurrent disease, and so on. We made improvements and the prevalence eventually fell but, as with most systems of intensive production, the attention to detail and the quality of installations and buildings needs to be consistently high to ensure that the problem is minimised. They took a few of the older buildings out of use and replaced them with new, better-insulated ones. Some people got the sack.

The prevalence of this problem has fallen, but when it was common anyone with a bit of inside knowledge could spot the lesions on the carcasses of affected broilers quite easily. I once came across a supermarket chiller where around two-thirds of the carcasses showed lesions of hock burn. I pointed this out to the manager. After a short discussion he realised I knew what I was talking about. I've never seen a chiller cabinet emptied so quickly.

Musculoskeletal conditions are common in broilers. There is growing societal concern that many broiler chickens have impaired locomotion or are even unable to walk. Despite considerable effort being expended to address the problems of leg health (for example, selecting strains with better conformation), the industry has been unable to eliminate the problems of lameness and leg deformity in the fastest-growing broilers. One study of 51,000 broilers showed that at a mean age of 40 days, over 27.6% of birds showed poor locomotion and 3.3% were barely able to walk, that is over 1,600 birds. The authors concluded that the primary risk factors associated with impaired locomotion and poor leg health were those specifically associated with rate of growth, and that any changes to the management to reduce leg health problems would be likely to reduce growth rate and production.[23] There does seem to be a strong genetic correlation between rapid growth and poor musculoskeletal health. In other words, selecting for better musculoskeletal health is, in effect, selecting for slower growth. Which is exactly opposite to the objectives of the industry.

Opinion is divided on whether poor welfare, including contact dermatitis, is a function of stocking density or whether other factors are more important. A 2010 review for the European Union concluded that many of the common pathologies of the broiler (breast blisters, chronic dermatitis and leg disorders) are a result of high stocking densities, and quoted studies showing that locomotor behaviour, preening and general activity are reduced and disturbance of resting is increased at higher densities.[24] A large-scale study (2.7 million birds) involving experimental manipulation of stocking density under a range of commercial conditions, however, concluded that differences among producers in the environment that they provide for chickens have more impact on welfare than stocking density itself.[25]

Despite this division of opinion, in 2007 the EU adopted conditions for the broilers that limited the stocking density to a maximum of 33 kg per square metre.[26] This can be increased to up to 39 kg per square metre and, in special cases, up to 42 kg per square metre, but extra requirements apply.[27] You might think that such detailed rules would be last the word, and this was clearly the intention of the EU. But there is pressure for further reform. Large retailers have been put under pressure to agree to the 'Better Chicken Commitment' – with some success.[28] Signatories commit to improvements in husbandry, to adopting more humane slaughter methods, and to third-party auditing - all of which should improve the lot of the broiler. This is an example of how citizens acting collectively can force improvements – a subject I'll return to in Chapter 12.

If there was any doubt about the welfare of broilers, there is a final twist. The hens that lay the eggs that hatch into broiler chicks are known as 'broiler breeders'. As you might expect, both sexes carry genes which confer upon the chick the ability to grow rapidly as soon as it hatches. This intensive selection for production traits, especially growth rate, is associated with an increased nutritional requirement and thus feed consumption. In essence, broiler breeders, like broilers, have prodigious appetites and a tendency to get fat and unproductive. To counter this, feed is severely restricted during rearing. This severe feed restriction has negative effects on bird welfare as it causes chronic stress resulting from hunger. Further, broiler breeder males may be aggressive during mating, which can cause severe injuries to the hens – therefore requiring routine mutilations such as beak trimming, de-spurring and toe clipping to reduce the risk of injury.[29] In short, even if the problems with broiler production were resolved, we would still be left with severe welfare problems associated with the parent birds.

In considering the impact of farming and other interventions on animals, I've questioned whether constraints on the expression of the normal repertoire of behaviour may affect welfare. It's logical to ask the same questions about broilers. Does intensive broiler rearing prevent the expression of normal behaviour? I think it's difficult to tell. Selective breeding has produced a docile, fast-growing and hungry bird, but because it is killed so young, territorial and breeding behaviour doesn't get a chance to develop. But there are enough other concerns about the rest of the system, particularly the physical effects of sustained fast growth, to make me consider carefully whether I can stomach this degree of exploitation.

My concern is that there is little to be done here other than to vote with your feet. The business model is based on producing a cheap product, which is achieved by maximising the amount of chicken meat per unit area, in the shortest possible time. And that involves what I have just described – extremely highly selected, fast-growing birds kept in a controlled environment at a high density and fed on a high-protein and energy-rich diet.

While production experts and scientists seek to resolve the welfare problems that this system generates, sometimes with success, there is little prospect of the industry adopting a model which allows the birds to grow more slowly, with fewer developmental problems. Given that the global demand for chicken meat is growing exponentially, I see little prospect of radical change. It comes down to this: if you believe the welfare implications are unacceptable, then don't buy chicken meat derived from birds reared in these conditions. Of course, I may be wrong, and the Better Chicken Commitment may permanently change things for the better. Let's hope so.

Is there an alternative? There is, but note that perhaps 98 out of 100 birds are reared in the intensive conditions I've just described. In alternative systems, diets are less protein- and energy-rich and the strains of birds used are slower-growing, with a less extreme muscular profile. Stocking densities are lower. The birds may be free range with access to the outside, although more typically they will be reared in sheds and provided with natural light and a less barren environment. The result is a bird that reaches a slaughter weight at anything between 8 and 14 weeks. It is a more mature animal and has bones which feel like bones rather than rubber, and it has a much firmer flesh which, unlike a standard broiler, tastes exactly like a chicken should. However, the combination of slower growth and lower stocking rate means fewer birds from a given area, with the result that an oven-ready chicken from an alternative system is between three and five times more expensive than a conventional broiler of similar weight. You pay your money and you take your choice.

Making these choices is difficult. I try to buy free-range broilers or where they are advertised as being less intensively reared, and to avoid the industrially produced birds whenever I can. Achieving the latter is difficult when eating out or when buying prepared meals. It's always worth asking – but be prepared for some blank looks.

Which comes first? It's the egg

The other main player in poultry production is the egg-laying hen, commonly known as a 'layer'. As with the broiler industry, the sheer scale of production is mind-boggling. At any one time, in Britain, there are estimated to be 38 million layers in production.[30] We consume around 30 million eggs per day, or 11 billion per year.[31] That equates to each of us consuming a little less than half an egg per day, or 165 per year. And that doesn't take account of imported eggs and egg products. Nor does it take account of people who don't eat eggs. If you are thinking 'I don't eat four eggs a week', just think of the other products that you eat that have eggs in them – cakes, egg fried rice, meringue, custard, and many others.

Like the broiler, the modern layer is a sophisticated hybrid bird developed over multiple generations to do one thing – in this case, to consume food and turn it into eggs. It does it rather well. It is a much more compact bird than a broiler. There is little point in having a large-bodied bird if the objective is egg production.

From hatching to about 22–24 weeks of age, the layer grows to a point at which it is mature enough to lay eggs. This is known as a 'point of lay' (POL) pullet. Provided it is well cared for, it will then begin to lay eggs continuously, at the rate of around one a day, for around 300 days. Although the potential lifespan of a laying hen is around six years, few commercial layers are allowed to live that long. This is because after a year's laying production drops off – the bird lays fewer, albeit larger eggs. In general, layers are considered 'spent' after 72 weeks of age or thereabouts and sent for slaughter (generally to be processed into pies and other meat products), to be replaced with a new group of POL birds, and the cycle of production continues.

The British consumer has an odd relationship with the egg and the laying hen. First, unlike most other markets, we crave brown rather than white eggs because they are considered 'healthier' despite the evidence that there is no nutritional difference. In most other markets, white eggs are the norm, and not without good reason – the white hens that lay white eggs are less aggressive, eat less food and lay more eggs on average than the brown hens which lay the brown eggs.[32] Second, unlike the rest of Europe, we have consistently opted to buy a far greater proportion of our eggs from free-range systems than from other systems such as those where the birds are caged.[33] But, as we will see, despite the widely held view that free-range systems provide better welfare for the laying hen, it isn't necessarily so.

Although entrepreneurs in the USA had experimented with keeping laying hens in cages during the 1930s, until the 1950s egg production in Britain was something of a cottage industry. Units would rarely hold more than 250 hens, and generally the birds would have access to range; that is, the opportunity to roam free and forage on pasture in an enclosure of a few hundred square metres. Diseases, particularly those caused by internal and external parasites, were a major problem. As was predation. While nest boxes were provided in the coops, a substantial proportion would be laid outside in nests built by the hens and either lost or found only after they had spoiled.

As post-war demand grew, innovators and early adopters became involved. Deep-litter systems where birds were housed but not otherwise confined provided benefits over free range such as less predation and cleaner eggs, but disease and so-called floor eggs remained persistent problems. The introduction of the battery cage system required a greater initial investment, but provided for intensification and mechanisation, as well as improving hygiene (particularly of the eggs) and the physical health status of the hens and reducing feed and labour costs.[34] At the same time, improvements in genetics, in disease control and in nutrition contributed to a steady increase in the number of eggs produced per bird each year.

The battery cage system confines laying hens to a small area and a barren environment and hence inhibits many of the normal patterns of behaviour that the birds are able to display when free-ranging such as nesting, foraging and dust bathing. The wire-mesh cages were generally arranged in three or four tiers, with each one holding between four and six birds. Each bird had a floor area less than that of an A4 sheet of paper. The floor was sloped to allow the eggs to roll away to the back of the cage to allow for easy collection. There were no nest boxes, dust baths or other 'furniture', and because of the confinement and the small space allowances, the birds were unable to flap their wings and perch normally. Houses would hold up to 20,000 birds.

It is worth noting that the introduction of the battery cage pre-dated the debate about behavioural needs of animals and, while there was some disquiet about the conditions in battery systems, for the most part people were grateful for the availability of cheap, clean and plentiful eggs. In effect, partly because of other priorities, behavioural needs were effectively trumped by the benefits of efficient production. By the 1970s, in the UK the vast majority of eggs were produced in battery systems.[35] However, the publication of Ruth Harrison's *Animal Machines* had stimulated a public debate,[36] and over the next three decades the arguments raged over whether the keeping of hens in cages was inhumane or not.

Eventually, at least in the European Union, a degree of scientific consensus emerged. A report of the main scientific advisory group of the EU concluded that 'It is clear that because of its small size and its barrenness, the battery cage as used at present has inherent severe disadvantages for the welfare of hens'.[37] In 1999, the European Council agreed a directive which banned the conventional battery cage in the EU from 2012, after a 12-year phase-out period. The same directive provided for detailed rules for the keeping of birds in free-range and other non-cage systems where the birds are permanently housed.

The move to furnished cages met more of the needs and wants of laying hens than the previous cage system. Studies had demonstrated that much of the enrichment that the new cages provided, such as nest boxes, perches and dust baths, were furnishings that the birds were prepared to 'work' for.[38] In other words, these substrates facilitate the expression of more of the birds' normal behavioural repertoire than hitherto. However, some animal welfare organisations, such as Compassion in World Farming (CIWF), continue to call for enriched cages to be prohibited on the grounds that they provide no significant or worthwhile welfare benefits compared with conventional battery cages. CIWF believes that good animal welfare depends on three components – physical wellbeing, mental wellbeing and natural living – and that all three of these are compromised by confinement in enriched cages.[39]

It is true that the extra space per bird is not great, and that the birds are still confined in what can only be described as a cage.

In 1991, at the height of the debate about the needs of the laying hen, I was privileged to study animal behaviour at Edinburgh University under the late Professor David Wood-Gush, an expert in the behaviour of the domestic chicken. He took the view that the quality of space provided for an animal was often more important than the quantity. Sadly, he died before the EU rules were agreed. I sometimes wonder whether he would have agreed with the EU's solution. It certainly improves the quality of space, but arguably not much else – the birds are still confined in a very small area.

In introducing a ban on 'barren' cages, and in taking into account the move to non-cage systems, it has been argued that legislation concerning laying hen welfare is influenced more by public perceptions than by scientific and commercial evidence. The usual objection to battery cages ignores the fact that they were developed in order to *improve* the health of hens, and that there are more advantages than disadvantages with cage systems in comparison with alternative (non-cage) housing. Why then, it was argued, does the public remain more concerned about just one of the Farm Animal Welfare Council's Five Freedoms – to display most normal patterns of behaviour – than about the other four?[40] When the consumer is offered the choice of eggs produced from cage systems and free range, it raises the question 'Is the customer always right?' As we will see, the choice between the two systems (not forgetting barn eggs – that is to say, eggs from housed birds but not kept in cages) is not clear-cut.

There are two other welfare concerns associated with layers that we should consider. The first is bone loss or osteoporosis. Commercial layer hens have been genetically selected to lay a very high number of eggs, and are consequently highly susceptible to poor bone strength and osteoporosis. Osteoporosis is related to nutritional imbalance, level of egg production, and the birds' inability to move and keep their muscles and bones healthy. The formation of egg shells requires the deposition of calcium. Since eggs are laid at a very high rate, this leads to a loss of bone calcium. Osteoporosis is an end result of this process and is a widespread problem in commercial hens.[41] Welfare implications of osteoporosis include pain, debility, and mortality associated with bone fracture.

Laying hens in battery cages have the poorest bone and muscle health, whereas hens in cage-free systems that promote physical activity, such as free range, tend to have less osteoporosis and better musculoskeletal health. However, hens in cage-free systems sustain a higher rate of fractures during the production period, probably caused by flying into objects such as perches and nest boxes. One study that investigated bone damage in laying hens in the UK found that 36% of hens from furnished cages had a keel bone (sternum) fracture, and this figure was even higher in non-cage systems, where prevalence levels were between 45% and 86%.[42] While hens in cage systems sustain fewer fractures during production, their bones are frequently broken during collection, handling and transport prior to slaughter. Studies show that the prevalence of bone fractures at the time of slaughter can reach around 30%.[43]

Furnished cages allow hens to perch, which contributes to improved bone and muscle strength and results in the lowest number of total fractures compared with both cage-free and battery cage systems. Yet it is clear that the genetics of the high-performing laying hen predisposes to the development of osteoporosis, and that it is a problem in all systems of production.[44]

The second welfare concern is feather pecking. Feather pecking is the pecking, pulling out and eating of feathers of conspecifics, and it is an important welfare issue in laying hens. Whether it occurs in conventional battery cage systems, in group-housing systems such as large furnished cages, or in non-cage systems including free range, it is difficult to control. While genetic, nutritional, management and social factors have all been implicated, no one intervention has been wholly successful at eliminating the risk of outbreaks. Instead, the industry relies on an intervention to reduce the likelihood of an outbreak starting.

The partial amputation of the beak at up to 10 days of age (beak trimming) is generally effective in reducing feather pecking – but there are obvious welfare implications when the sensitive tissues of such an important organ are disrupted. The beak is a complex, functional organ with an extensive nerve supply, including nociceptors that sense pain and noxious stimuli. In his book *Bird Sense*, the ornithologist Tim Birkhead describes the ability of many species of bird when foraging to distinguish between similar objects, for instance a piece of gravel and a morsel of grain of the same size, weight and shape. Because of the way the bird's eyes are set in the head, the bird does not see the object or the end of its beak but uses its senses of touch and taste, both of which are closely associated with the beak.[45] In a sense the bird's beak provides the same sensory feedback that we get from the tips of our fingers when we feel for things in the dark. You can imagine how unpleasant and how difficult life would become if your fingers were somehow physically altered. While the nature and extent of beak trimming has become substantially less invasive as the former widely used technique of cautery has been replaced with infrared technology, it is still a mutilation, and a number of European states have introduced a prohibition or are actively considering one.

The UK government has stopped short of a prohibition. Defra's *Code of Practice for the Welfare for Laying Hens and Pullets* makes it clear that beak trimming should be stopped 'as soon as reasonably possible'.[46] In the meantime, the arguments continue between the poultry sector, which believes that beak trimming is an intervention that protects welfare, and the animal welfare organisations, which consider it an avoidable mutilation.[47]

What next for the egg-laying hen?

Like large-scale broiler production, large-scale egg production comes with inherent welfare problems. Although a number of improvements have been brought about by government intervention and industry efforts, many of these problems remain or have been only partially resolved. Working on the assumption that you want to continue to eat at least some eggs, which system provides the best welfare for the birds? As you might have guessed, there is no easy answer.

There will always be a compromise between the maintenance of health and hygiene on the one hand, and the expression of normal patterns of behaviour on the other. Battery systems, whether enriched or not, are capable of producing a more hygienic product with lower labour and feed costs, have much lower levels of disease and parasites, and virtually eliminate predation. However, the system suffers the serious disadvantage that the ability of the hen to express the normal range of behaviours is inhibited (though somewhat less so when the cages are enriched). In contrast, free-range systems allow for the full range of normal behaviour but have the disadvantages of a high prevalence of disease, greater aggression and an inability to control stocking density at critical points. In summary, the debate about which of the systems best meets the various welfare needs of the laying hen is far from over. In reality, because of the scale of production required to meet consumer demand, perhaps no commercial system is capable of meeting welfare needs entirely.

Keeping hens on smallholdings where the birds have plenty of room to forage, dust bathe and nest, where disease can be controlled effectively and there is suitable shelter and protection from predation sounds like the best approach. Group sizes can be kept small, which helps to maintain a semblance of how the birds would once have lived in the wild. Remember the small social group size of the ancestral form of the laying hen, the red jungle fowl. Individual recognition between birds seems to be limited to groups of up to about 80 birds,[48] so in much larger groups it may not be possible. Further, in battery cages small groups show higher production levels and decreased aggression, hysteria and other behavioural problems compared with larger unit sizes.[49]

But let's be realistic. Little flocks are not going to supply each of us with 160 eggs a year. So how do you make a choice? In 1991 two welfare scientists attempted to rank the various systems according to their welfare benefits.[50] They concluded that 'It is evident that the relationship between the welfare of laying hens and their environment is complex, and that choices between husbandry systems are difficult, especially as economics have to be taken into account.' However, they went on to say that the ideal system for present-day strains of laying hens should provide for small group sizes together with freedom of movement and complex environment characteristics. 'That's free range,' I hear you say. Well ... it is, sort of, but it is not quite that simple. The authors did not preclude confining birds: small group sizes in a complex environment can still be housed and even in a cage. It depends on the size of the cage – in practice there is little difference between a large cage and a small building.

The EU Directive sets no upper limit to the number of hens that may be kept in a group that are described as free range, provided detailed rules about housing and the density of birds on the outdoor range are adhered to.[51] Even in the relatively rare circumstances where a maximum free-range group size has been set, the upper limit is generous. For example, under the EU organic standards it's 3,000 hens, and the Soil Association organic standards specify a maximum flock size of 2,000.[52]

Despite concerns about the welfare of free-range laying hens, the British market continues to grow. Eggs produced from hens in free-range conditions accounted for 63% of all eggs produced in 2021. Several supermarkets have announced that by a given date they will no longer sell eggs from cage systems.

What of the future? As with broiler systems, I can't see much prospect of significant change. As systems become globalised, governments worry that independently adopted welfare rules will put domestic producers at a competitive disadvantage against foreign producers. They are therefore unlikely to act unilaterally, if at all, particularly as the EU appears to believe the problem has been resolved via the detailed rules agreed in 1999.

I avoid eggs from caged systems but, for the reasons set out above, I am sceptical about the welfare of free-range birds, especially at the upper ends of the scale. And as for pre-prepared egg pasta, who knows what type of egg it contains? The alternative is to seek out local producers who farm sympathetically. Go and have a look or, better still, keep your own and look after your flock to ensure that welfare standards are high. And you can always make your own pasta.

Turkeys and ducks

In comparison to chickens, turkeys and ducks make up a small but non-trivial proportion of British poultry production. There are around 14 million of each species slaughtered for food annually.[53]

The wild turkey is native to North America. Despite adult birds weighing between 4 and 11 kg, they are agile, fast fliers. They are omnivorous, foraging on the ground or climbing into shrubs and small trees to feed. They are social animals, and will form seasonal flocks of 10–20 birds.

Farmed turkeys suffer from welfare problems similar to those that beset broiler chickens. Modern commercial turkeys have been selectively bred for fast growth and disproportionately large breast muscles. They are slaughtered when they are between 9 and 24 weeks of age, and may weigh upwards of 20 kg. A high proportion of turkeys are kept in intensive indoor systems with no outdoor access. These systems restrict natural behaviour, although smaller, seasonal operations do often provide access to the outside. Selective breeding has also limited their behaviour. Far from being able to fly, farmed turkeys often have difficulty walking and mating; they are often too heavy to mate naturally, and instead females are usually artificially inseminated. Keeping turkeys in barren environments increases aggression between the birds, which leads to stress and injuries. To reduce aggression such as feather pecking and cannibalism, turkeys are often kept in low lighting conditions to make them less active. And as with laying hens, turkeys have their beaks trimmed when they are one day old, to reduce injury from feather pecking.

The ancestral form of the vast majority of farmed ducks is the mallard, a familiar, hardy and prolific species. Like most ducks it spends a great deal of time on the water, feeding, socialising and roosting.

The conditions in which ducks are reared are quite similar to those of broiler chickens, although ducks tend to be stocked at a much lower density and, possibly because they are not as fast-growing as broilers, ducks have fewer musculoskeletal problems. It is not surprising then that many people assume that farmed ducks have permanent access to water – but that is neither the law nor convention for intensively reared ducks, other than for drinking. Animal welfare organisations such as the RSPCA and CIWF argue that ducks must be given access to open water. A detailed review by the RSPCA concluded that the welfare of ducks is compromised by denying access to open water,[54] citing clear evidence that 'duck welfare is related to the nature and extent of their access to water',[55] and that 'ducks have a behavioural need for freely accessible open water'.[56]

Farmed fish

Obviously, fish are completely different to pigs and poultry. However, fish are farmed with exactly the same objective in mind as the other species. That is, to produce the maximum amount of marketable meat from a given area (or, in the case of fish, a given volume of water).

Farming fish is now big business in Britain. Over 90% of farmed fish production is dedicated to the Atlantic salmon. In 2019, over 200,000 tonnes of salmon were produced in the UK, mainly in Scotland, representing about 8% of global salmon production.[57] This is just a drop in the ocean of total global fish farming production, however, which is estimated to be 178 million tonnes per annum (2019).[58]

Atlantic salmon are reared through various stages in cages tethered in sea lochs. They are fed a carefully controlled diet consisting mainly of processed sea fish with small amounts of grain and animal by-products. They grow quickly and have a very low feed conversion ratio – often as low as 1.1–1.3, which means that at certain stages an Atlantic salmon will eat 1.1 kg of food and increase by 1 kg in weight. In these respects, they are, then, even more efficient than broilers. It's easy to see why fish farming is such an attractive prospect.

There is an important difference between farmed fish and other farmed animals, however. Fish require substantial amounts of high-quality protein if they are to grow fast. This means that, unlike poultry and pigs, most farmed fish have a diet that contains a high proportion of fish meal (although there are studies under way on replacing some of this with protein from plant sources).

The percentage of protein in a broiler diet varies between 19% and 23%, whereas in farmed fish it is 30–55%. Formulating such a high-quality diet requires large volumes of low-value sea fish, such as sand eels, to be hoovered up from the sea and compounded into pellets to be fed to the farmed fish, to produce high-value protein in the form of salmon steaks. This profligacy contrasts starkly with the ability of the ruminant to convert a resource that is of little direct value to people (grass) to high-quality protein (beef and lamb). The impact on fish stocks and hence on fish to catch for market, as well as on the abundance of other species that rely on such fish, such as breeding seabirds, is difficult to tease out from other factors – but it is seems unlikely to be a positive one.

The use of open cages for rearing Atlantic salmon in Scottish sea lochs is controversial, but the impact on the immediate environment is clear. Excess food and fish waste increase the levels of nutrients in the water and have the potential to lead to oxygen-deprived waters that stress aquatic life. The use of chemicals such as antibiotics, antifoulants and pesticides has unintended consequences for marine organisms. Farmed fish also appear to threaten the wild population: viruses and parasites transfer between farmed and wild fish, and escaped farmed salmon can compete with wild fish and interbreed with local wild stocks of the same population, altering the overall pool of genetic diversity.

According to a Farm Animal Welfare Committee (FAWC) 2014 report on farmed fish welfare, the most important factor that affected salmon welfare was similar to that affecting the welfare of broilers: the immediate environment, which in the case of the fish means the quality of the water in which they are reared. The FAWC report stated that more work is needed on aspects of water quality such as carbon dioxide levels and water temperature.[59]

CIWF believes that the behavioural requirements of most of the fish species used in aquaculture are poorly understood. It is unlikely that the conditions in intensive farming meet even the basic needs of fish. It is beyond dispute that overcrowded fish are more susceptible to disease and suffer more stress, aggression, and physical injuries such as fin damage. Along with lack of space, overcrowding can also lead to poor water quality and oxygen depletion. This is the view of CIWF, who seek substantial reform of fish farming.[60] However, FAWC concluded that, on its own, stocking density is not necessarily one of the most important things affecting fish welfare. Perhaps counterintuitively, some species of fish seem to prefer a higher density. Stocking densities that are too low can cause certain species of fish to become territorial and aggressive. Arctic char, for instance, show fewer agonistic behaviours and higher growth rates at high densities. Similarly, aggressive interactions were negatively related to stocking density in Atlantic salmon, and their overall welfare score was best at an intermediate density, indicating that both low and high densities may affect welfare.[61]

Studies generally use absolute stocking densities based on total biomass, which seldom reflect the actual fish density inside the cage, as fish choose to congregate at specific depths depending on environmental conditions. Position in the water column may affect disease susceptibility. A particularly high-profile problem affecting Atlantic salmon is the sea-louse, an external copepod parasite, which, when infestations are large, can cause severe ulceration and even death, generally through increasing the susceptibility to other disease. There appears to be no sustainable treatment regime: chemical treatments damage the environment, and even the use of cleaner fish – introduced to nibble the lice off the skin of the salmon – has problems as the larger salmon prey on the smaller cleaners. Worse, the cleaner fish, generally species of wrasse, have to be caught and transported long distances. And although there is evidence that the extent of the problem is partially density-dependent, a low stocking density does not appear to reduce the risk of infection.[62]

Fish farming and other forms of aquaculture are relatively new but are growing at a phenomenal rate. More and more species are being exploited: the proposed establishment of an octopus farm in Spain is controversial, not least because of the increasing evidence of the intelligence and sentience of octopuses.

The needs of each species will be different, and it is clear that for farmed fish, given the diversity of species worldwide and the relative novelty of rearing most of these species at scale, there is much still to be learned about the impact of intensive production on fish welfare. It would be wrong, given the differences, to simply extrapolate from what we know about birds and mammals and attempt to define their needs on that basis. However, many of the welfare issues raised by FAWC in addition to water quality, such as physical damage especially to fins and skin, disease, parasites and impacts on social behaviour, are similar to those that beset intensively farmed mammals and birds. And as FAWC stated, given that fish are able to detect and respond to noxious stimuli, it is important that 'deliberations on management and other processes should be made on this basis'.

Many of the welfare concerns that appear to affect farmed Atlantic salmon are not fully understood. Taking into account the environmental impacts of the most commonly used systems, one would like to think there would be caution about further expansion. On the contrary, production in Scotland rose fourfold between 2010 and 2020 and is projected to increase by another 50% by 2030. In another parallel with broiler meat, salmon steaks are now plentiful and cheap, the price having risen more slowly than inflation over the last 20 years. Arguably, like broiler meat, as well as being cheap, farmed salmon steaks are now bland, uniform and tasteless.

Perhaps the lack of public interest in the plight of the farmed salmon is similar to the lack of interest in the poisoned rat: out of sight, out of mind. And (perhaps) we also take the view that 'it's only a fish'. Some time ago, I took the decision not to eat farmed salmon for two reasons: the environmental impact and my concerns about poor welfare.

Summary

Each of the species covered in this chapter lends itself to intensive rearing. They are fast-growing and reproduce rapidly, need relatively little space and, while their diets are expensive in comparison to grass, they provide a steady return on investment. However, this comes at a price – maximising growth and production requires confinement of one sort or another so that, while disease is often well controlled, the full repertoire of the animal's behaviour is frequently severely limited, to the extent that its physical and mental wellbeing suffer.

This is the price we pay for relatively cheap animal products. However, there are alternatives – lower-input and lower-output systems tend to take greater account of the wellbeing of the animal. Outputs are lower per unit area, and generally animals take longer to reach marketable size and weight. The products are more expensive as a consequence.

For the concerned citizen this is simply a matter of choice. And unless we all adopt a vegan diet, the choice is stark: either cheap meat or better animal welfare.

6

Snares, Guns and Poison: the 'Management' of Wildlife

I started this book by claiming that most animals are exploited by us, in one way or another. Is this also true for wildlife? In contrast to pets, farmed animals and research animals, wildlife is unconfined and is free to live and die without a helping or interfering hand from us. Why then devote a whole chapter to wildlife, in a book about the exploitation of animals? It is simply this – because we manipulate the landscape and the animals that inhabit that landscape to such an extent that those activities can be considered exploitation. It is very rare to find any wildlife that is free of the effects of human presence and meddling. Some might have escaped our ministrations in the more remote parts of the globe, Antarctica for example, but there is none to be found in Britain, or in most of the developed world.

A self-confessed wildlife nut

I've been watching wildlife since I was a teenager. I wore the badge of the Young Ornithologists Club with pride, and for over 50 years I have kept a tally of the birds I've seen in Britain. I still go 'birding' almost weekly and I've taken up photography with a few dozen published photos to my name. When work took me overseas, I packed binoculars and a field guide and generally found a way to visit national parks on my days off – seeing everything from manatees in the Caribbean to alligators in Florida to platypus in Australia. In recent years, I've had wonderful wildlife experiences in Japan and South Africa.

There are some marvellous wildlife spectacles in Britain. I've been lucky enough to experience huge flocks of pink-footed geese in Aberdeenshire, the gannetry on the Bass Rock and a superpod of common dolphins in the Atlantic. Closer to home, I have seen murmurations of starlings on the

Somerset Levels and Moors and all six species of native reptile in a single weekend.

In contrast to many places overseas, where I admit I was spoiled, there are few places where wildlife is abundant in Britain. Some remote seabird colonies contain staggering numbers of birds, and some of the winter goose and wader roosts on the coast really have to be seen to appreciate them. But aside from a few intensively managed reserves there is really nowhere inland where large numbers of any native species can be seen.

Britain has a severely depleted fauna – both in numbers and in diversity. Why is that? Much of this depletion is attributed to habitat loss, of which more later. But there is also another cause – the deliberate and systematic intervention against wildlife, lethal or otherwise, where the species in question is deemed to be a 'problem'. It is clear that some species are persecuted or have been persecuted to the point where their very existence, locally or nationally, is under threat. Some species are beginning to bounce back after decades of persecution – the pine marten and the polecat, for instance. A more enlightened attitude has certainly helped here. It's unfortunate and very concerning that similar protections can't be applied to other species of mustelid like the stoat and the weasel, the populations of which are poorly understood. And if that isn't of sufficient concern, many of the methods to kill birds and mammals deemed to be a 'problem' appear to be indiscriminate and inhumane. My concerns, which are detailed in this chapter, are therefore twofold – loss of diversity and inhumane equipment and practices.

How do we exploit wildlife?

The vast majority of wildlife is subject to some sort of human intervention. This includes varying amounts of population control, confinement behind fences, recreational shooting and hunting, 'pest' and predator control (including trapping and poisoning), deliberate and accidental extirpations, releases and reintroductions, habitat management and conservation activities such as tagging, bird ringing, wildlife rescue and rehabilitation. The built environment has an impact on wildlife that includes everything from the loss of habitat when land is developed, to bird strikes on buildings, and the effects of transportation such as roadkill and the fragmentation of habitats.

Is all of this exploitation? It's clear that the harvesting of wildlife for food, with or without interventions to reduce predator numbers or to provide optimal habitat for the quarry, is a form of exploitation. It is clear to me that the killing of rodents and other wildlife that are conveniently and often carelessly dumped in the 'pest' category is also a form of exploitation. But if I accidentally run over and kill a rabbit during my drive to work, or if a blackbird dies after flying into my conservatory, is that exploitation? It might not be direct or deliberate, but given that the deaths of the rabbit and the blackbird are direct consequences of our manipulation of the environment, these are impacts that should concern us all.

Does this mean we are obliged to protect wildlife in all circumstances and from all harms? I don't believe so. Most factors affecting the welfare of wild animals – such as predation, the weather and disease – are outside of our direct control, and it is neither practicable nor desirable to influence the welfare of wild animals in all circumstances.

Free-living wild animals will suffer and die, and it is generally inappropriate to interfere with the natural course of events. Given the efforts expended by conservationists, gamekeepers and government to manipulate the environment and its wildlife, it is clear that such a view might not be universally held. There is a balance to be struck, and, along with efforts to ensure abundance and diversity, we need to consider how anthropogenic activities affect the welfare of wildlife while remaining aware that a decision to intervene is often influenced by personal values.

If we conclude that wildlife doesn't need to be protected from itself and the elements, why worry about the effects of human intervention? Remember 'a rat is a rat is a rat' – regardless of where it is located and what level of care and protection we provide, the rat has the same capacity to suffer. While artificial and arbitrary groupings of animals (research animals, farmed animals, pets, etc.) might be convenient when seeking to regulate activity or to justify exploitation, they can be unhelpful when challenging the orthodoxy of the types of exploitation we have tolerated hitherto.

I argue that we owe the same consideration to wildlife that we give to farmed animals and research animals. I'm not naive: I don't expect the standards that apply in a slaughterhouse to apply to the killing of wildlife. But, when it is necessary, I expect wildlife to be killed humanely by competent people who are publicly accountable. When a decision is made, for example, to kill a fox because it is eating your chickens, there is a moral imperative to follow best practice. I appreciate that this might seem blindingly obvious to the proponents of shooting and 'pest' control, and those seeking to protect their interests, but it isn't always so. There are a few things that need to be done before reaching for the shotgun or setting that snare. First, are you certain that it's a fox that's the culprit? Second, have you taken reasonable steps to prevent the fox from gaining access to the chickens? Third, are you sure the means of killing is as humane as is possible? In essence, as well asking 'how?' we need to ask 'why?'

Some people will be outraged by this advice. Some will believe that their interventions against wildlife already meet these tests. They don't need advice from an interfering third party. At the other extreme, some people will be outraged that I am even prepared to countenance the killing of a fox, under any circumstances. Leaving that argument to one side for now, if it has to be done, it has to be done humanely. Sadly, all too often that's not the case. Regardless of the rationale, many of the ways in which wildlife is killed and trapped are demonstrably inhumane. For the remaining methods, there is often insufficient evidence for us to be satisfied that it is humane. There are many reasons for this, and much of this chapter is devoted to exploring these concerns more fully.

Of course, there are many ways in which we intervene, directly or indirectly, in the environment that affects the fate and experience of wildlife. Everything from excluding animals from our houses, farm buildings and crops, to bird scaring, to lethal control. Much of it is relatively benign and doesn't involve killing or harming wildlife. However, much that appears benign and humane is neither – and it pays to consider these in a little more detail.

How do our activities affect wildlife?

For centuries, humans have altered the landscape and its associated flora and fauna. There is good evidence that, as humans spread out of their native Africa and colonised Eurasia, Australasia and the Americas, the numbers and diversity of large mammals and (mainly flightless) birds fell dramatically. Driven over cliffs and speared for food and skins, many species became extinct within a few centuries of human colonisation. Although climate change may have played a part, most authorities believe that human 'overkill' contributed to the extinction of megafauna such as ground sloths, moas and mammoths.

Once settled agriculture developed and became the norm, the manipulation of the landscape proceeded apace: from felling trees to tilling the land, from the herding of animals to the draining of wetlands. Over the last two or three millennia, the development of the built environment changed the landscape still more dramatically. The impacts are not always obvious or immediate, but the shaping of the landscape and the impacts on wildlife have a long history.

Changes to the environment will be to the detriment of some species, while others may benefit. For example, the skylark, a bird of open country, will benefit from the opening up of grasslands as woodlands are felled, but the draining of wetlands means the loss of prime redshank breeding habitat.

More direct interventions might benefit one species to the detriment of another. That may indeed be the express intention – to kill a proportion of one species, say the red fox, to enhance the population of another that is deemed important, such as the non-native ring-necked pheasant. Similarly, a conservation organisation may kill carrion crows to improve the breeding success of the threatened Eurasian curlew. The removal of scrub on a nature reserve might enhance the grassland that supports the food plant for the endangered large blue butterfly, but the loss of dense cover reduces the amount of suitable breeding habitat for the whitethroat and blackcap.

Agriculture introduces another dimension. Wildlife may start to feast on the growing crops, but when scaring and other deterrents fail, measures to kill the marauders are developed and deployed – traps, guns and poisons – of which more later. Keeping herbivorous animals often attracts predatory animals – and, again, when deterrence and secure fencing fails, farmers may resort to killing. And for some, pretty soon this becomes the norm.

Businesses, particularly any involving the storage and preparation of food, from a grain store to a restaurant, are likely to be attractive to rodents. 'Pest' control involves a number of interventions from trapping to poisons, providing businesses both with an assurance that their interests are protected and demonstrating to customers and regulators that best practice is being followed. Similarly, householders who find evidence of mice or rats in their kitchen will frequently resort to traps and poison in an attempt to eliminate them.

Although Chapters 3, 4 and 5 investigated the exploitation of farmed animals, the impact of farming on wildlife was not covered. This has been and continues to be profound. An investigation into the impacts of farming and other land uses on the diversity and abundance of native wildlife in Britain is beyond the scope of this book, and the subject is well covered elsewhere.[1] There are also excellent books which investigate the impact of habitat loss on wildlife and how diversity and abundance can be restored.[2] Here, I will focus on interventions that affect individual wild animals more directly

Interventions involving wildlife fall into five broad categories:

1   Protecting birds as quarry for shooting – 'game' birds
2   Protecting other interests such as farming and fisheries
3   Protecting businesses and households from 'pest' species
4   Killing wildlife to conserve other species or habitats
5   Killing wildlife for food and recreation

The remainder of this chapter will examine the first three of these categories. In Chapter 7, the activities of conservation bodies are examined, and this will include the fourth category above. Killing wildlife for food and recreation is covered in Chapter 8.

These categories are helpful but they are not hard-and-fast groupings. There is an overlap between killing for food and killing to protect business interests; the snaring of a rabbit for food and the snaring of a fox that is killing pheasant poults use similar techniques and equipment. Similarly, the shotgun used for grouse shooting is similar to that used to kill woodpigeons. And, perhaps most importantly, there is likely to be little or no difference between the experience of the rabbit shot for the pot and the rabbit shot because it is munching its way through your cabbages.

How does the law protect wildlife from harm?

The law that protects wildlife is complex and subject to frequent changes. Each nation of the UK has its own laws, each subtly different, further deepening that complexity. While the law protects the welfare of farmed animals, pets and research animals through imposing a duty of care on the person in charge of the animal, for obvious reasons no such duty of care exists for wildlife – except when it comes under human control.

Animal welfare legislation such as the Animal Welfare Act 2006, the Animal Health and Welfare (Scotland) Act 2006 and the Welfare of Animals Act (Northern Ireland) 2011 is primarily intended to protect domestic animals. Non-domesticated animals are also protected under these Acts when they are under the control of humans or not living independently in the wild. This is generally taken as including animals in traps.[3] The Wild Mammals Protection Act 1996 makes it an offence to carry out a variety of acts against wild mammals including mutilating, nailing, impaling and crushing, any of which might feasibly occur in a spring trap. Again, however, the offence depends on there being an intention to cause unnecessary suffering. And, in addition to the usual and sensible exemption to allow killing when an animal is in extremis (for example, one that is badly injured and likely to die), the Act does not apply when using otherwise legal methods of capture and killing such as traps and poisons.[4] As we will see, this is convenient, since without this exemption many of the routine means of capture and killing of mammals and birds would be illegal.

Other legislation provides further protection. The Wildlife and Countryside Act 1981 prohibits certain methods of killing or taking wild animals and provides full protection for certain mammals and (as a starting point) for all birds.[5] In some cases this is further enhanced by additional legislation – for example, the Protection of Badgers Act 1992.[6]

Curiously, while all birds are protected unless licences are issued allowing for one form of intervention or another, for mammals (and other vertebrates and some invertebrates) there is no such blanket protection. So except for deer and some highly protected species such as bats, badgers, pine marten and polecat, protection is very limited. For instance, there is no restriction on when you may kill foxes, rabbits or brown hares, and minimal controls over the methods.

There are two main types of licence. These can be issued for birds and mammals, although the circumstances vary considerably:

- A **General Licence** allows for interventions that would otherwise be illegal and is issued when it is believed that the activities licensed 'carry a low risk to the conservation or welfare of a protected species'.[7] You do not need to apply for a General Licence but you must follow its conditions of use. Carrion crows, certain other corvids and a small number of other species may be trapped and killed or shot under licence. Other General Licences are issued for mammals but these are more limited. As we will see, given the very low level of scrutiny of compliance with these licences it is difficult to see how a conclusion about the welfare impacts can be reached. Following legal challenge in 2019 by the organisation Wild Justice (wildjustice.org.uk), the list of species covered by the General Licences has been reduced and the conditions changed, in effect making the routine killing of birds covered by the licences subject to more stringent conditions.

- An **Individual Licence** (sometimes known as a specific licence) might be issued, on application, to an individual or an organisation to take, kill or disturb a protected species not covered by the General Licence and where evidence is provided that mitigating measures, such as scaring, have failed or are impractical.[8] For example, when a bird begins to nest in a food processing factory and it cannot be removed without disturbing the nest or harming the chicks.

Is the law effective in protecting wildlife from harm? Clearly the law does not prevent the killing of all wildlife. Nor is that the intention. But does the law seek to ensure that it is done humanely? Let's examine this.

Do we know whether wildlife is killed humanely? Evidence from the badger cull

There are few published studies about the humaneness of the methods used for killing wildlife in Britain. There are a number of reasons for this. First, there is little official recording of the methodology and the numbers involved; second, regulation is generally 'light touch'; and, third, the proponents tend to be reticent about their activities. It follows that much of the information about wildlife welfare derives from anecdotal reporting or is inferred from indirect studies rather than from direct observations.

There is one data set, derived from a study into the licensed killing of badgers, which sets the benchmark for assessing the humaneness of killing wildlife. Before examining the evidence about other interventions, it is worth looking at this in some detail.

The badger is one of the most heavily protected species of large mammal in Britain, but, despite this protection, it is also the subject of one of the most widespread officially sanctioned killing programmes of any large mammal in the developed world. The badger is blamed for the maintenance and spread of bovine tuberculosis, a disease that besets the British national cattle herd. To counter this threat, since 2013 the government has authorised several dozen private companies – founded specifically for the purpose – to kill badgers in England. This is not a trivial matter. Currently, the area where badger killing is licensed is larger than the combined areas of the Kruger, Royal Chitwan and Everglades National Parks. The government does not publish a running total but reports indicate that over 100,000 badgers have been shot and killed since 2013,[9] with over 35,000 killed in 2020 alone.[10] In the absence of an exit strategy, there is no end in sight.

Attempts to eradicate the infection from cattle in Britain began over 100 years ago, and despite protracted efforts and enormous public expenditure, the infection is still spreading to new areas. This, despite over eight years of systematic killing of our native wildlife and increasingly stringent controls on cattle.

It is important to acknowledge that this case study is possible only because there has been independent analysis of the killing of badgers. This is in stark contrast to the vast majority of the killing of other wildlife. The data and supporting analysis provide answers to a number of questions which I would suggest could usefully be posed for all wildlife killing.

Is the method of killing humane?

As a basic principle, a wild animal killed for any reason should, where possible, be rendered instantaneously insensible to pain and remain so until death ensues.

When, in 2013, the government decided to pilot the systematic killing of badgers, considerable effort was put into ensuring it was done humanely. Detailed protocols were developed with experts in firearms and the shooting of mammals. These included a number of options: the shooting badgers at night with a rifle or a shotgun ('controlled shooting'), or live trapping followed by shooting. An expert group, the Independent Expert Panel (IEP), was appointed to oversee the collection and analysis of data on the humaneness, effectiveness and safety of badger killing. The report is one of the most comprehensive available on killing wildlife, because it takes into account the fate and experience of all targeted animals, including those shot at and not recovered.

From the pilot study, the IEP concluded: 'We have very high confidence that robust monitoring protocols were put in place to monitor the effectiveness and humaneness ... We have very high confidence in the data collection and analysis performed by [the government].' However, the IEP expressed concerns about the welfare of so-called controlled shooting:

> Evidence suggests that between 7.4% and 22.8% of badgers that were shot at were still alive after 5 min and therefore were at risk of experiencing marked pain … If culling is continued … standards of effectiveness and humaneness must be improved. Continuation of monitoring, of both effectiveness and humaneness, is necessary to demonstrate that improvements have been achieved.[11]

Despite the conclusion that 'If culling is continued … standards of effectiveness and humaneness must be improved', since 2014 the non-retrieval rate (NRR) in shootings monitored by Natural England has consistently been between 9.5% and 13.8% (though it is interesting to note that the NRR reported by the cull companies themselves averages a little over 2%).[12]

Despite the IEP's advice, the amount of monitoring remains below the level at which one can be confident that killing is humane. The percentage of controlled or free shooting observed by Natural England staff was between 8% and 20% in the first three years of the cull, but since 2017 it has been consistently below 1%.[13] Determining the humaneness of shooting badgers at night is difficult. Using a third party to observe shooting at a substantially higher rate than has been the case would provide better data. Examining a proportion of badger carcasses to determine the wound tract caused by the rifle bullet would provide evidence of accuracy. A representative sample would allow a credible estimate of how many badgers had been shot in accordance with best practice. None of this is done in numbers high enough to allow meaningful conclusions to be drawn.

Because of concerns about shooting badgers with a rifle, the British Veterinary Association has argued that trapping followed by shooting at close range is likely to be more humane.[14] Despite this, the vast majority of badgers are shot with a rifle, in the dark, at a distance of up to 70 metres.

Is the method justified?

Defra and Natural England publish an Annual Monitoring Summary which contains a statement about humaneness. This is from the 2020 summary:

> The likelihood of suffering in badgers is comparable with the range of outcomes reported when other control activities, currently accepted by society, have been assessed.[15]

There are two things worth noting about this statement:

- First, studies on the efficiency and humaneness of shooting free-living wild mammals are scant, and it is therefore impossible to meaningfully compare badger shooting with the shooting of other species.

- Second, the phrase 'currently accepted by society' needs a closer look. We are expected to rely on some notion that badger killing is 'no worse than killing other animals' and that society is comfortable with the method and outcome. This bothers me for two reasons: (1) with the possible exception of stalked deer, there are few or no data on the humaneness of the killing of other wildlife, and (2) it is quite a stretch to assume that the public are comfortable with the 'other control activities' for the simple reason that I can find no evidence that the public has ever been asked.

Does the citizen get any say?

Decisions are made in secret. Licences are granted to private companies; the name of the company, its directors and its accounts are all kept secret. When new cull areas are considered, only certain residents are approached. In practice, small tracts of land not considered material to the success of the operation are rarely brought in. Other residents such as householders, even when they have badger setts on their land, are neither consulted nor informed.

Unlike other activities that make substantial changes to the local environment, such as building new roads or 'fracking', local communities are excluded from decision making around badger culling. The locations of the areas licensed for killing and the start and end dates of each killing season are also kept secret.

The government argues that secrecy is necessary because of the perceived threat of disruption by those opposed to badger killing. The secrecy and lack of engagement continues despite rulings from the Information Commissioner that the potential for disruption should not override the requirement to publish information about the culling programme.[16]

* * *

In conclusion, officially licensed killing of badgers cannot be said to follow good practice, because the means by which the majority of badgers are killed is not the most humane method, the shooters are inadequately monitored, and there are inadequate levels of scrutiny of the cull as it proceeds.

We will return to official killing of badgers in Chapter 12 when we consider the application of ethical principles to wildlife 'management'. But before that we need to look at the other situations where wildlife is routinely killed.

Protecting birds as quarry for shooting – so-called game birds

Much of what we deem wildlife, while it might be free-living and unconstrained (or, in the case of pheasants, unconstrained after being artificially incubated, reared in cages after hatching, treated for parasites, fed for several weeks in a 'release pen' and then released), resides in an environment that has often been manipulated to maximise numbers, whether it be for shooting or for conservation. And the killing of animals and birds deemed a threat to the objectives of those involved is, on some tracts of land and to suit certain objectives, simply part of an annual, unchanging routine of activity.

There are several legal ways in which birds and mammals deemed to be 'pests' can be killed. But what about the methods of killing? Is it humane? How is it regulated? Is there independent scrutiny? These questions apply, of course, to both the shooting of the quarry and the killing of 'pests'. To be clear, while the killing of 'pest' species occurs on shooting estates, farms and land managed for conservation, much also takes place in business premises and houses. The circumstances when animals are killed to aid conservation of other species are covered in Chapter 7.

Considerable effort goes into protecting precious species of 'game' birds (I try to avoid using the term 'game' since it perpetuates an arbitrary categorisation of animals – but I employ the term here simply because it is widely used). This involves killing a substantial but unknown number of a variety of species of mammals and birds that are deemed to be a threat to the shootable surplus of birds produced by the business. Let's not forget that all this killing is to protect three main quarry species, two of which are not native to Britain – the ring-necked pheasant and the red-legged partridge – and one of which is essentially farmed, albeit in extensive conditions – the red grouse. Much smaller numbers of a native species, the grey partridge, are also protected for shooting.

The methods of 'pest' control involve not only shooting but also lethal traps, live capture and snares. The evidence base that supports most of these trapping methods and the equipment that is in routine use is at best, incomplete, and at worst, makes a good case for prohibiting their continued use. But don't take my word for it. Take a look at the evidence. It's worth considering these methods in some detail, while bearing in mind that their use is not confined to protecting birds for shooting.

Lethal traps

These fall into two categories: regulated spring traps, which in Britain are generally required to meet particular welfare approval standards, and unregulated spring traps.

Regulated spring traps are spring-loaded, with the mechanism triggered by the arrival of the target animal in the trap. Tripping of the trap releases a mechanism which applies rapid force to the target animal with the intention of causing the rapid onset of unconsciousness and death. These are mainly used for killing small to medium-sized mammals such as rodents, and stoats

111

and weasels – all of which are considered to be a threat to 'game' birds. Other mustelids such as the pine marten and the polecat are protected, though that doesn't always save them. Spring traps are also used for killing grey squirrels and rabbits. Some more modern traps (e.g. Goodnature traps – https:// goodnaturetraps.co.uk) are powered by pressurised carbon dioxide gas and designed to re-set automatically.

Traps are generally sited in locations frequented by the target animal, and placed and set in a way that is either attractive to the target species (a spring trap baited with food, for example) or on a regular 'run' where the target animal is likely to pass through or over the trap. There is no requirement for the periodic inspection of lethal traps. Spring traps must be deployed with a tunnel or a similar adaptation to reduce the likelihood that any larger non-target species will be caught. Recent changes to legislation require traps used for stoats to meet the Agreement on International Humane Trapping Standards (AIHTS).[17]

Unregulated spring traps of the type designed to kill mice and rats are generally known as break-back traps or snap traps. There are similar doubts about the humaneness of these traps, but unregulated spring traps are rarely used to kill the animals which are believed to threaten 'game' birds.

The use of spring traps is widespread, largely unscrutinised by the regulators, and not subject to any competence requirement on the part of the operative. Anecdotally, spring traps are frequently poorly and illegally deployed and frequently catch species other than the intended quarry.[18]

Live capture

Live capture in traps is a common method to catch members of the crow family (Corvidae) including the rook, jackdaw, carrion crow, hooded crow, jay and magpie. These birds are variously considered to represent a threat to wildlife, 'game' birds, crops and, less commonly, other property. There are a number of designs, including Larsen, ladder and similar traps.[19] They consist of a frame, wooden or metal, with a wire mesh covering. All types operate by attracting the target species into the trap, either with a decoy bird of the same species or by placing food in the trap. Depending on the trap design, once the bird enters a particular part of the trap, either it triggers a mechanism which will tip it into the interior or it will find its way into a chamber from which it cannot escape.

These traps must be operated under conditions of General Licences issued by Natural England and their counterparts in Wales, Scotland and Northern Ireland. This includes checking the traps at a frequency of at least once every 24 hours and the provision of water, food and shelter for the decoy bird. In Scotland there is a requirement for each trap to be individually identified and for each to be registered with NatureScot, the regulator.[20]

There are no requirements which take account of their behavioural needs, the stress of confinement or the forced close and protracted proximity with congeners and other species, including predators. There is a requirement to kill the birds humanely, but there is evidence that capture and ultimately killing is a highly stressful experience.[21] Non-target species, including pro-tected species such as birds of prey, are also caught and, even if released alive, may suffer and potentially die as a result of entrapment. The welfare of the decoy birds is also likely to be adversely affected by captivity and the inability to behave naturally.

Corvids are intelligent, resourceful and highly social animals. In the absence of evidence to the contrary, and taking into account the evidence of the highly developed cognitive abilities of the corvid, it is safe to assume that even if the physical needs of the captured birds are met, the capture and confinement of such highly intelligent birds can only have a detrimental effect on their welfare.[22] Take into account the impact on the welfare of decoy birds, that the operatives are not obliged to demonstrate competence in humane killing and the unknown but substantial number of non-target species caught, and the case for better regulation or even prohibition becomes unarguable.

Snares

A snare is an anchored cable or wire noose set to catch wild animals such as foxes and rabbits. In Britain, snares must be 'free-running' so that they can relax once an animal stops pulling, thereby allowing the operative to decide whether to kill the animal or release it unharmed. Swivels on snares are required. The free end of the wire must be firmly attached to a post or rock sufficient to prevent an animal pulling the trap and itself away from the trap site; dragging (non-fixed) anchors are prohibited. Once set, a snare must be inspected at frequencies of no more than 24 hours.

In contrast to the situation in some other countries, snares are widely used in Britain to restrain animals for despatch rather than as killing devices. Self-locking snares, which can act only with a one-way ratchet effect to tighten around the neck as the animal pulls, are now illegal in Britain (although no clear legal definition of 'self-locking' has been established).[23] In 2011, the Scottish government tightened the legislation governing the use of snares, and introduced a requirement for an identification number on all snares and the keeping of detailed records.[24] Further changes to the law in Scotland introduced compulsory training and a statutory five-yearly review of the use of snares.[25] Curiously, but perhaps unsurprisingly, these reviews concentrate solely on refining practices and the law about the use of snares, rather than any consideration of whether snares ought to be used in the first place. Many European countries have banned the use of snares (as well as Larsen and similar types of traps) completely, although I have found little information about how such decisions were reached.

There are very few data on the welfare impact of snares. Opinion is divided. On the one hand, an independent group reporting to Defra in 2005 believed that, if they are used carefully, their adverse welfare consequences can be relatively minor.[26] Another report, albeit commissioned by an animal welfare organisation, asserts that 'snares have long been known to inflict extreme physical and mental suffering on captured animals'.[27]

Snares capture a wide range of non-target animals, including protected species such as badgers and otters, as well as dogs and cats. There is no doubt that if used carelessly or irresponsibly (and especially if not frequently inspected, or if an animal escapes while still entangled) they can cause severe welfare problems. Clearly much depends on operator practice. Because there are no adequate data it is not currently possible to assess the welfare impact of snares under routine use or how frequently severe problems occur. There is almost no information about the welfare impacts, or rates of non-target capture, associated with the setting of snares to catch rabbits.[28]

The lack of data available on the use of snares, and particularly on their welfare impact, is a serious problem both for making assessments about when the use of snares is justifiable and also for developing guidelines about good practice.[29] However, even when a detailed Code of Practice (CoP) is developed with advice from practitioners,[30] there is evidence that compliance is poor. Observations of fox snare users showed that none was fully compliant with the CoP. For instance, most were using non-compliant snares and many were failing to avoid sites where the snare could become entangled with nearby objects, such as branches, trees or fences. Similar observations were made when accompanying professional rabbit snare operators. Many operatives were using non-compliant but freely available proprietary snares.[31]

The internet is full of lurid and distressing photos of animals caught in snares, often with horrific injuries. While there is no reason to doubt that with the very best practice, as in scientific studies, snares can cause little physical harm, I find it very hard to conclude that, on balance, the benefits outweigh the harms, potential or actual. The idea that it is acceptable to arrest a running animal by the neck with a wire noose and leave it to struggle for as long as 24 hours without water, food or shelter is to me, whatever the potential benefits, an unacceptable practice. When one considers that a substantial but unknown number of animals are seriously injured in the process, both target and non-target species, it becomes doubly unacceptable.

Despite these misgivings and continuous campaigning for a ban, nothing much has happened. Shamefully, few of the recommendations in the 2005 independent report to Defra for further study have been followed up and substantial data gaps remain. A recent review for the charity Animal Aid by Professor Stephen Harris concluded 'the available data show that it is impossible to monitor the use of snares, or enforce legal requirements, Codes of Practice or best-practice guidelines', adding that 'the only way to stop extremely high levels of non-target captures, illegal use and misuse of snares, address animal welfare concerns, and recognise that wild animals are sentient beings, is to prohibit the use of snares.'[32]

* * *

In summary, large numbers of native and non-native mammals and birds are killed using techniques and equipment that are demonstrably inhumane, to protect the interests of businesses that rely on the largely uncontrolled and indiscriminate release of industrial quantities of non-native birds. Add to that the evidence that many of the birds die before being 'presented to the gun' and that substantial numbers of those that are shot go to waste, and the argument for continuing without substantial reform is pretty thin.

The law governing control methods is complex, out of date and has a poor evidence base. One might expect that, given the growing interest in animal welfare, it would be for the proponents of these methods to amass the evidence of humaneness, or to devise humane alternatives. However, in practice, it falls to those campaigning against their use to advance evidence of the harms caused.

In addition to the legal methods of killing native animals that are deemed inconvenient, there are two other things we should be concerned about: the use of illegal methods to kill target species, and the illegal killing of protected species. I am as outraged as the next person about these activities, and I can only hope that those campaigning for better enforcement are successful. However, alongside this, I'd like to see greater and closer scrutiny of the legal practices. The lack of official scrutiny is indefensible, since even these legal methods of killing animals are clearly the cause of substantial suffering.

Protecting farming

Farming of all types has a significant impact on wildlife. Habitat change wrought by intensive farming (along with the built environment) reduces the abundance and diversity of wildlife. Despite that, farmland proves an ideal environment for certain species. Some species of deer, badgers, rabbits, brown hares and red foxes can be very common on farmland. Around farm buildings mice and brown rats can be abundant. Although moles are often a woodland species, they can become widespread on both livestock and arable farms.

Each of these species has an impact on farms. Deer can hinder the growth of farm woodlands and forestry by browsing the growing saplings, badgers can undermine roadways and damage crops, rabbits, hares and woodpigeons cause considerable losses to arable crops and horticulture.

The mole: how not to 'manage' wildlife

Of all of the species controlled on farmland and elsewhere, one rarely seen species is worth investigating in more detail: the mole. This is because the way the mole is treated is an example of all of the worst aspects of our interventions against wildlife.

The European mole (*Talpa europaea*) is one member of a family of tunnelling mammals which spends most of their time underground. Its preferred natural habitat is woodland but, as we often see from the presence of their molehills, moles seem quite content in pastureland, household lawns, golf courses and even bowling greens. They feed on invertebrates, primarily earthworms.

The appearance of moles and their molehills on carefully tended patches of grass can be enough to drive the custodians to apoplexy, although the extent of the damage they do is contentious. Certainly, the activity of moles tends to be more obvious in the spring when males are seeking mates and defending territories. Once they mate, there is less movement between territories and hence fewer new tunnels and fewer molehills. That information may not be of much comfort to the golf-course greenkeeper in April looking at a slew of new molehills on the greens.

There are all manner of substances and gadgets which are claimed to deter moles, including ultrasound devices and chilli powder. Results are, at best, mixed, and many people resort to lethal means. Poisons such as the previously widely used strychnine are now illegal and no longer available.

There are a number of different types of traps. All rely on catching the mole as it passes through an underground mole run. The two main types of mole spring trap used in Britain today are scissor (pincer) traps and Duffus (half-barrel) traps. Both are designed to catch moles around the body when a trigger plate or wire is pushed, releasing a spring-loaded mechanism that is intended to kill the mole by crushing. They are widely claimed by mole trappers to be humane.

None of the traps used for moles has been approved under the legislation as humane (nor are they required to be). Mole traps, like many other lethal traps, are freely available for sale. Have a look the next time you visit a garden centre. The traps come with a label attached that is typically about 40 mm square. I have seen more detailed instructions on a screwdriver.

People trapping moles need neither to be specially trained nor to demonstrate appropriate competences. There is no requirement for the periodic inspection of traps, although, in England, a voluntary code of practice suggests that traps should be inspected every 24 hours.[33]

Studies of the injuries caused by commonly used mole traps have been extremely consistent, indicating that no trapped moles died immediately as a result of damage to the cranium or vertebral column, with the primary identifiable cause of death in all but one case being acute haemorrhage, and most likely associated haemorrhagic shock. In a minority of cases, asphyxiation may have been the primary cause of death, but this may not be detectable at post mortem (although asphyxiation could potentially be inferred depending on other findings at post mortem).[34]

In summary, moles are routinely killed using traps that are exempted from an approval process that would otherwise provide some assurance that the trap was capable of killing humanely. The traps are freely available and there is no training or registration required for anyone wishing to use them. There is evidence from field studies that the traps are inhumane and probably cause a lingering death.

Let's remind ourselves: the mole is a mammal with a nervous system laid out in a similar fashion to humans, dogs and cattle and with all the same elements in its brain, albeit in different proportions, that provide me with the capacity to suffer, feel and anticipate. I can't prove that moles have the capacity to suffer, but I challenge anyone to provide evidence that they don't. And yet, through ignorance or indifference, we allow these animals to be trapped underground by means which are demonstrably ineffective and are probably the cause of prolonged suffering. Of course, this argument might be applied to many of the methods and species I've already described.

No one is held to account because, as the traps and the people using them are unregulated, there is no one who can be held to account. This, despite the results of detailed farm surveys that show moles do not constitute an agricultural 'pest' of any great importance.[35]

And there you have it. The mole is provided with no protection. A sentient mammal is condemned to a lingering death because, not to put too fine a point on it, no one cares.

Live traps are used for rabbits and foxes (as well as American mink and grey squirrels). Bait may be used to lure the animal into the trap, but live mammals or birds cannot be used. It is certain that any animals caught in such traps will suffer some degree of stress due to the confinement. In most cases, it is the intention ultimately to kill the trapped animal. Licenses may be obtained where badgers damage roadways and buildings but, unlike for bovine tuberculosis control, they are issued to deter and translocate rather than to kill. It is worth noting, however, that release of an animal elsewhere is not necessarily a humane thing to do – translocated animals may fail to adapt to or integrate into new territory and may suffer and die as a result.[36]

There are a number of other approved methods. Gas may be used to kill rabbits underground, but official advice is that less toxic alternatives such as shooting or trapping should be used in the first instance. Aluminium phosphide can be used only by professionals, for the fumigation of burrows. In contact with moisture, it produces phosphine, which is a highly toxic gas. It causes death by blocking cellular metabolism, leading to death through cardiac and respiratory failure. Affected rodents show signs of respiratory irritation and other forms of discomfort. There is no closed season for these species, which means that dependent young will starve when suckling females are killed.

Some birds, primarily corvids (carrion crow, hooded crow, raven, rook, jackdaw and magpie) are considered by many farmers to pose a threat to livestock, crops and nesting birds (as well as to 'game' birds – see above). Control is almost always lethal despite the requirement in the applicable licences to have exhausted non-lethal methods before killing birds. The methods used are shooting or trapping (similar to that detailed above). There is no closed season, and when adult birds are killed the nestlings will starve.

Other than the necessary certificate and licence for shotgun and other firearms, there are no competence requirements for any of these methods of killing. There is often no requirement for reporting or recording, although some of the more recent General Licences issued in Scotland do require some reporting. Individual (or specific) Licences generally include a reporting requirement.

Protecting fisheries

Another area of wildlife 'management' that is worth examining concerns the effect of various birds on recreational inland fisheries and commercial fisheries. The species involved are great cormorant, grey heron, goosander and red-breasted merganser. All are medium or large fish-eating species considered by anglers to have a detrimental effect on fish stocks. The evidence is often contentious.

There's no doubt that each species eats fish. What is at issue is the effect that the presence of these birds on a river or a stocked lake has on the numbers available to catch. Anglers have long argued that cormorants are damaging fish stocks, and that their numbers must be controlled.[37] There is evidence that management actions to reduce local cormorant colonies may have a positive effect on local perch abundance, but killing cormorants or reducing colony size has variable or weak effects on adjacent local fish of other species.[38] Another study, however, concluded that even though cormorants caught a lot of fish, ultimately fish numbers and biomass were little affected.[39]

A meta-analysis of the effect of several cormorant species on fish populations around the world concluded that the combined effect of cormorant predation on fish was negative, but the overall effect was not statistically significant.[40] Despite this equivocal evidence, several hundreds or thousands of cormorants are killed in England annually.[41] Although there have been calls for the cormorant to be added to the General Licences (that is, similar to carrion crows), only Individual Licences are issued.

Smaller numbers of goosanders and red-breasted mergansers are killed, ostensibly to protect salmon stocks – but, again, there is a paucity of evidence of the impacts of these birds on inland fisheries.[42] Again, control is through the issuing of Individual Licences, and the conditions that must be observed are considerable.

As with farmers' concerns about the impact of mammalian predators on their stock, one cannot help but think that the other factors that affect fish stocks such as climate change, pollution and aquaculture are more important than the impact of a few birds. But perhaps there is tendency to look for obvious scapegoats to blame for the dwindling catch. The cormorant is big, black and brazen and hence seems to fit the bill perfectly.

As with most of the unfortunate creatures covered in this chapter, when killing is authorised, provided the shooters have the necessary firearms certificate, then there is no requirement to demonstrate competence – either in shooting accuracy or in the identification of the species you are licensed to kill.

It is clear, despite numerous studies, that the evidence that these birds have direct and lasting impact on fish stocks remains equivocal. These species, while not threatened, should not be targeted without sufficient evidence; the burden of proof should rest with the advocates of their killing.

Protecting businesses and households from 'pest' species

It is highly likely that, at some stage, mice will appear where you live or work. Less frequently, you may be visited by rats. This is especially common in older and less well-maintained houses where rodents find it easier to get in. Many businesses, particularly food businesses, are at similar risk, and when this happens, proprietors are legally obliged to eliminate or exclude them.

In the late 1980s, we lived in a Victorian schoolmaster's house in rural Aberdeenshire. It was in a pretty poor state and as draughty as a pole barn. Each autumn, as the barley harvest was completed, numerous wood mice would migrate from the stubble to our house. I would find them helping themselves to the cat food or simply scurrying about the kitchen at night.

I started to use a snap trap (or break-back trap). One evening while we were watching TV in the next room, we heard a 'snap' and there was a dead mouse in the trap that I had set next to the cat-food cupboard. I disposed of the carcass and re-set the trap. The rest of the evening was punctuated by repeated 'snaps', and as the mice appeared to be smaller and smaller, I began to wonder whether the younger mice were venturing out to look for their parents who had failed to return. After the sixth dead mouse, I thought 'I can't be doing with this. There has to be a better way.'

I bought a trap advertised as 'humane' because it claimed to catch the mouse without killing it. It did. I caught several more mice alive and let them go in the garden. Each of them scuttled off (other than the one caught by my cat in its first moments of freedom), presumably to re-enter the house through the same hole they had used a few days before. My frustration with this method wasn't helped by a colleague who gleefully suggested that perhaps confining a mouse in a 'humane' trap for several hours was likely to be more inhumane that an effective snap trap that kills in a very short time. And dead mice don't go back into the house. I was back to square one. What about poisons? Deterrents? Or simple exclusion?

This then is the 'problem' of wildlife 'management' as experienced by the managers of shoots, farmers, conservationists and other landowners but in miniature. How to deal with wildlife that appears to be a hindrance.

Throughout this book I've advocated an ethical approach to all types of animal exploitation, including wildlife interventions. I've argued that an artificial and arbitrary distinction between different species depending on their 'use' and circumstances is wrong, and that we need to remember that the capacity for suffering is likely to be universal in the species of animal targeted. Which brings me back to mice in houses. Society, I believe, should expect that a consistent and evidence-based approach is taken to wildlife whether it's in your house, in a bistro, on your farm or on a nature reserve.

There are a number of ways of approaching the problem of rodents in a house, or anywhere else where they might be attracted. The most effective method is to prevent rodents getting into the building. Without access to food, water, shelter or nesting sites the house will not be attractive and they will go elsewhere. Food and food waste needs to be inaccessible. While modern houses are less easily accessed than older houses, it pays to look at all possible routes of entry in any house and block them. For instance, wire wool can be used to block holes and cracks, especially around pipes and places where the walls have been drilled.

Now though, let's assume that despite your best efforts, filling in the holes hasn't worked and mice and rats are still getting into the house. What options are open to you if you want to use a humane method? There are surprisingly few, and perhaps none of them meet the requirement for humaneness in full.

If you decide to use a lethal trap, unless you invest in one of the types that has been through the approval process, you will be using what is known as a snap trap – similar to what I was using in Aberdeenshire all those years ago. Traps of this general design (spring traps) are subject to official approval to determine whether they are capable of killing the quarry within a prescribed time. However, there is an exemption from approval for traps used for rodents and moles. Most of the snap traps used for trapping mice and rats that are purchased from high-street shops are the non-approved exempt type.

The snap trap is a simple device with a heavily spring-loaded bar and a trip to release it. It is generally baited with a morsel of food and it is the act of pulling the food from the trap which trips it. These traps are intended to cause death by crushing vital organs. Death can be very rapid, but if the animal is caught instead by a limb or its tail, for example, it will suffer pain and distress. Unless it is discovered promptly this may lead to a long drawn-out death.

Snap traps are widely used in households and businesses because they are cheap and freely available, but they require more labour than the alternatives such as rodenticides and glue traps. There are many different types, and larger sizes are used for rats. There is no type approval for any sort of snap trap, and no requirement for periodic inspection.

No type approval means there is no requirement for traps to be tested for humaneness prior to being placed on the market. Studies into a variety of commonly available snap traps show that performance is patchy.[43] While the direct welfare impact of the traps was not assessed, the wide variation in mechanical performance for each species, overlap in performance between rat and mouse traps, and increasing availability of weaker plastic rodent traps indicated that there is considerable scope for improving humaneness. In other words, because of inconsistencies in manufacturing, you cannot be confident that any of the commercially available traps are capable of killing a mouse or a rat humanely.

An alternative is the glue board. And yes, it is just as nasty as it sounds. Any small mammal (or bird) that attempts to cross the board or that alights on the surface will become permanently stuck to the glued surface. These traps are generally used in enclosed buildings such as houses and food premises. Glue boards are freely available and easily purchased online. The adverts often claim that they are 'safe, non-toxic and harmless'. Safe, that is, provided you are not a mouse, rat or other small creature that is unlucky enough to come into contact with one.

Glue boards are designed as restraining rather than killing devices, but the welfare of trapped target rodents and non-target animals is severely compromised and they often die as a result of thirst, starvation or their injuries rather than being humanely killed. Animals caught on glue boards often struggle to free themselves, and they are known to pull out their own hair, break bones and tear skin as a result. As a captured animal struggles, it may become increasingly stuck to the glue, and they sometimes drown in the glue as a result. Trapped animals may also become prey to other animals.

Industry guidance recommends that boards are checked at least every 12 hours.[44] However, substantial suffering can occur from the moment of capture. Glue boards may legally be used by members of the public (as well as by 'pest' control professionals), and even if trapped animals are discovered quickly, the untrained user is left with the problem of how to kill the rodents humanely. The more you examine these things the more you realise how horrible they are. That conclusion was reached in New Zealand several years ago after detailed consideration of the matter, and the use of glue boards was prohibited there in 2015.[45] Britain lags behind. It was not until 2002 that England and Wales introduced legislation to ban the use of glue traps. Similar legislation in Scotland remains under consideration.[46] The BVA supports a ban except for use by 'pest' control professionals.[47]

This brings us to poisons. Rodenticides or 'rat poisons' are the most widely used approach to the control of rats and mice. There are several types. The most commonly used are anticoagulants. These kill by disrupting blood clotting mechanisms and leading to death by blood loss. Bleeding can occur externally or into the gut, tissues, body cavities, joints, and inside the skull. Bleeding into joint spaces and inside the skull is known to be very painful in humans, and there is every reason to believe that anticoagulants have the same effect in rodents.[48] In 1997 the government's own regulatory body described anticoagulant rodenticides as 'markedly inhumane'.[49] Alphachloralose is another rodenticide which can be used to kill mice, though it is not legally permitted or suitable for use against rats. Because of its mode of action – anaesthesia leading to death through respiratory failure and hypothermia – it is thought to be more humane than anticoagulants.

Rats and mice can be caught in live-capture traps, the type I used after I'd become uncomfortable about killing mice with a snap trap. However, live traps have two important disadvantages. First, the impact on the welfare of the trapped animal confined for several hours is likely to be high. Second, you have to either release the captured rodent or kill it. Where do you release it? If the mouse is set free just outside the house then it will very likely come straight back in, unless you can find the route of entry and block it. If you take it some distance away or exclude it from your house there is a very real risk it will fail to adapt to a new territory and may suffer and die as a result.

Alternatively, if you decide to kill captured rodents, you need to be competent. This is not just a matter of ethics, it is the law: the Animal Welfare Act 2006 applies to captive rodents as it does to any other wild animal under human control. Not only do you need to kill the animal humanely, but whether you intend to release or kill it you must inspect the trap frequently enough to avoid the animal starving, becoming water-deprived or succumbing to exposure.

On the face of it dealing with rodents in your house is simple and straight-forward. All you need to do is put down a few packs of user-friendly mouse poison or a few snap traps and it will soon be sorted. However, as soon as you take into account the risk of animal suffering then it becomes much more complicated.

Some way of determining the relative welfare impacts of trapping and baiting methods would be useful. A recent study on killing rats reports exactly that – but it does not make comfortable reading because none of the available methods could be described as consistently humane.[50] The greatest welfare impacts were associated with poison baiting methods and with capture on a glue trap, followed by concussive killing; all of these methods should be considered last resorts from a welfare perspective. Lower impacts were associated with cage trapping, followed by concussive killing. The impact of snap trapping was highly variable. The authors concluded that snap traps should be regulated and tested to identify those that cause rapid uncon-sciousness, and that these might offer the most humane method.

While I accept that it is quite a challenge to achieve, making the house less attractive (for example, keeping food securely) and making it less easy to get into is probably a better strategy. There is an excellent practical guide to humane rodent (and mole) control on the website of the Universities Federation for Animal Welfare.[51]

The inhumaneness of most rodent killing highlights an interesting paradox in the way we treat different classes of animal. Animals killed for food, research or their fur are never legally permitted to suffer for hours, let alone for days. Indeed in laboratories, slaughterhouses and veterinary practices, acceptable killing methods usually have to act in seconds. And yet we tolerate the most ghastly treatment of rodents.

Decision making and accountability

When most wildlife is killed, regardless of the reasons for the killing, humaneness appears not to be a major consideration. There are exceptions, such as the detailed rules for killing of badgers and the obligation, for certain species, to use traps approved for the purpose. However, even these measures are of limited value since the shooting of badgers is so poorly monitored and there is evidence that approved traps are ineffective. This contrasts starkly with efforts to protect farmed and research animals.

Decisions about why and how to kill wildlife need to be taken carefully. The evidence about intelligence, sentience and the capacity to suffer grows inexorably. And then there is the evidence of our own eyes. If you've ever been lucky enough to watch groups of crows, ravens, jackdaws and choughs cavorting and tumbling through the air then you'll know that these are highly intelligent sentient animals. The acrobatics appear to serve no direct purpose – it might be play or a display to boost an individual's status. I've concluded it is simply *joie de vivre*, and until there is cast-iron evidence to the contrary, I'm sticking to that view.

It is likely that crows are more intelligent than dogs, and yet we treat or allow others to treat these birds abysmally. Why? I think because most of us are ill-informed and some of us simply don't care. Those who routinely trap and kill them often act through tradition or prejudice, and because outdated legislation allows it.

The objectives of wildlife killing should be specific, measurable, and outcome-based, where the outcome relates to the desired reduction of harm – such as reducing crop loss, preventing disease transmission, or furthering the conservation of a priority species – rather than simply reducing the number of target animals.[52] Equally important is the need for the killing to be done humanely. In other words, we need to apply sentiments and standards similar to those that we follow when we kill animals for food.

However, as this chapter has shown, there are many ways in which the killing of wildlife fails to meet these basic criteria. We should care about this because if these animals have the capacity to suffer, like farmed animals and research animals, we have a moral responsibility to ensure that they are treated with care and respect. It might be argued that the existing legal provisions are sufficient to protect the welfare of wild animals, but given the almost complete lack of monitoring and enforcement, there are no means of determining whether or not this is true.

It is clear that much trapping (including snares and live traps) is inhumane. And because their use is not seasonally limited, snares and traps capture lactating and pregnant animals or juveniles, including within populations of protected species that may be adversely affected by the use of non-selective traps.

It is also clear that there is insufficient accountability about decision making. Large numbers of animals are killed, yet the decisions are made by a small number of people and with little or no scrutiny by regulators or society as a whole.

I have every sympathy with campaigns against trapping, snaring and the use of Larsen and similar traps. I am similarly outraged at the scale and methods of badger killing. I am infuriated at the indiscriminate killing of native wildlife to protect industrial quantities of non-native birds for shooting. And don't get me started on moles. Railing against these iniquities, however, just doesn't seem to achieve a great deal. We get fobbed off with platitudes and protracted reviews – which somehow, despite the earnest efforts of those involved, seem to change little.

I propose that we reset our entire approach to wildlife. That might be best done by articulating how we will value wildlife. It is not something which exists simply as a resource, the numbers of which can be dialled up or down according to some arbitrary objective. Wildlife is part of our shared heritage and contributes to the richness of the environment. As well as maintaining and enhancing the abundance and diversity of our native wildlife, however we exploit those animals, whether it is for food, for our entertainment or for our viewing pleasure, necessary interventions must be justified, sustainable and above all humane.

A decision by the Scottish government in 2020 to set up an advisory body comprising experts on animal welfare may herald substantial change. The Scottish Animal Welfare Commission will focus on protecting wild and companion animals while also providing scientific and ethical advice to government. An advisory body with animal welfare in its remit is not new – the Farm Animal Welfare Committee dates back to the 1970s. But a body set up to advise government on the welfare of wild animals is new. This offers the prospect of resetting the approach to wildlife welfare, and we should watch developments with interest – particularly those parts of its workplan that address wildlife welfare.

An important part of this reset is a set of principles to govern most, perhaps all, circumstances when we intervene against wildlife. I believe it should include the following:

- A requirement to consider and document what alternative means of control have been considered and why these have been rejected as ineffective.
- A requirement for training and accreditation to ensure that operatives are demonstrably competent and have knowledge of the ecology of target species.
- A decision to intervene that should not rest solely with the occupier of the land. The citizen needs a say.
- The setting of a threshold of 'harm' in statute which describes the circumstances when different methods of control can be used – for example, damage to property, crops or livestock, or impacts on species of conservation concern.
- A requirement to record or report any activities associated with control, including the impact on populations of target and non-target species.
- A requirement to assess the humaneness of all current methods of lethal control. If evidence cannot be provided, then the methods must be withdrawn. No new methods to be introduced unless demonstrably humane.

- The deployment of traps and other methods of killing to be limited so as to avoid leaving dependent young.
- An effective inspectorate that ensures compliance.

Finally, let's ditch the notion that killing an animal can be justified simply because it is abundant. The fact that badgers are common and widespread should not provide the authorities and landowners with a free pass for licensed killing. Such a position is founded on the mistaken belief that 'another few won't hurt'. Evidence of the benefit has to be unequivocal. There are two reasons for this. First, the experience of an animal targeted for killing is no different whether it is abundant or close to extinction – it has the same capacity to suffer. Second, the inference from such a rationale is that we should only value and hence protect a particular species when it is threatened. There are many examples where the killing of large numbers of animals proceeded unchecked for too long, eventually leading to extinction of the species. Even small-scale declines reduce the chances for all of us to enjoy watching wildlife in our local countryside. These are not 'animal rights' arguments but simply a plea to adopt a broader consideration of the issues when making decisions about whether to intervene against wildlife.

We will return to these principles in Chapter 12 when we consider, in the context of a wider ethic for animals, both the ethics of wildlife intervention and how society could and should play a bigger part in decisions that affect our wildlife.

7

Conservation: Exploitation with Clear Limits?

When I left full-time work six years ago, I started volunteering for the RSPB and Natural England. Most of this is survey work, such as counting breeding waders on the Somerset Levels and Moors or looking for snipe nests in meadows before hay cutting starts. Occasionally, in the summer I get the opportunity to do something a bit more immediate and 'hands-on' – I catch baby cranes. I'm not a licensed ringer but I'm part of a team which catches and rings Eurasian crane chicks as part of the Great Crane Project – the reintroduction project in Somerset. The objective is to catch the eight-week-old chicks in the few days before they fledge. Any earlier and the rings and transmitter will be too big for the growing bird. Much later and the chicks will have fledged and be impossible to catch. This means a carefully planned operation.

In Somerset, the cranes generally nest on open, flat ground, which means they can see you coming from some distance away. So we go out at night, and it goes like this: at around three in the morning, led by the RSPB's estimable Damon Bridge, five of us meet about 2 kilometres from where the crane family was seen to go to roost the evening before. Then, in near-silence and in the pitch black, we walk to the field where the family has been feeding over the last few days. We are each taken to a predetermined spot on one or other of the damp field boundaries and told by Damon to keep our heads down and sit tight.

There I lie for what seems hours but is really only about 90 minutes. As the sun rises, sedge warblers begin to sing and I can just make out grey herons flying overhead. Shapes appear from the gloom and gradually I see where I am – perched on the edge of a ditch in a four-hectare field. Abrupt messages from Damon begin to appear on my phone ('The birds have left the roost', 'The birds are walking to the field') and then, once it is properly light, we get the all-important message: 'The birds are in the field. Jump up and run!' So we do. After the cold and stiffening ditch, getting up and running, even for an old codger like me, feels good.

As we run to the centre of the field, the adult cranes take to the air in panic. The chick, still unable to fly, disappears. While the adults circle around us, calling anxiously to the chick and warning it to stay hidden, we spread out, quartering the field, hunting for the unfledged chick. After about five minutes, I find a scrawny, long-legged bundle of skin, bones and feathers clinging to the ground, trying to look as small as possible. Alison, our licensed ringer, takes over. She catches and restrains the bird, skilfully avoiding the sharp bill and the long legs.

Next, the young crane is ringed. As well as the uniquely numbered BTO metal ring, Eurasian cranes have three colour-coded plastic rings applied to each leg. These provide a country code, the year of hatching, and a unique identifier allowing each bird to be identified at a distance. One of the rings has a radio transmitter built into it. Normally a sock would be put over the bird's head to calm it while the job is done, just like a blanket plonked over a budgie's cage to help it settle down for the night. Alison has forgotten to bring a sock, so on this occasion, I am proud to report, the crane chick wears my bobble hat. Once it has been ringed, Alison collects two downy feathers to allow laboratory DNA-based sexing. The whole process, from hearing the shout to letting the chick go, takes between 5 and 10 minutes.

When it is released, the chick lunges at our legs several times with its sharp bill, but then it wanders off, seemingly oblivious, back to its parents. As I watch the chick retreat, two things spring to mind. First, given that capture and handling are likely to be very stressful, are there any measurable impacts, for example, on bird mortality? Second, what do we know about the physical impacts of rings and tags on birds?

Do conservation organisations exploit wildlife?

Nature conservation organisations exist, primarily, to maintain and increase wildlife and the habitat upon which it depends. Of course, they may also have secondary objectives such as educating the public and providing access to open spaces but their *raison d'être* is conservation. Many of these organisations set out to protect and enhance wildlife without a monetary or other metric which defines success. For instance, the Wildlife Trusts assert: 'We believe that a healthy wildlife-rich natural world is valuable in its own right.'[1]

You might ask, 'If conservation NGOs are dedicated to biodiversity and increasing abundance they can't possibly be involved in animal exploitation, can they?'

Remember, though, that any intervention unless it is trivial can be considered as a form of exploitation. But even if we accept that, how, you might wonder, could any of their activities have a negative effect on animal welfare? In the vast majority of cases, they don't. But there are a number of conservation activities where animal welfare may indeed be compromised. These vary from apparently benign interventions to collect data on animals of interest, such as the ringing or tagging of animals, to efforts to improve the numbers and range of certain species by translocating them from one place to another.

Less benign interventions include actions such as predator control to protect vulnerable species, habitat manipulation to enhance the breeding success of threatened species, and denying access to predatory animals, through fencing for example, to areas reserved for priority species. To be clear, none of these activities are comparable to many of the other interventions carried out by the other sectors under scrutiny in this book. But that does not mean that conservation NGOs get a free pass.

If we wish to hold these other sectors to account, conservation has to be held to the same standard. Indeed, there is an argument that in promoting themselves as contributing to saving the natural world and the animals that inhabit it, conservation bodies need to be held to a higher standard. In other words, they need to be exemplars of best practice.

Conservation NGOs seek to improve diversity and abundance in the small areas of land which they control. But increasing diversity and abundance, particularly for threatened species with particular habitat requirements, requires management. For certain objectives, this involves the removal of some animals – for instance the culling of deer to help protect the woodland understorey that is essential for breeding nightingales and other woodland birds.

Most conservation NGOs are membership organisations. Members are often motivated by a love of nature, and the idea that the organisation that they have joined is killing the animals that they believed their subscriptions are there to protect does not always go down well. There is a balance to be struck and a fine line to tread in order to achieve conservation objectives without compromising animal welfare or alienating members.

The need for transparency

Conservation and animal welfare do not always sit well together. Part of this stems from a view that while the collection of certain types of data (from bird ringing, for example) is not without risk to animal welfare, its utility is so great that the ends justify the means. Similarly, it may be argued that the killing of an abundant species in order to protect one that is more vulnerable is beyond challenge. I don't believe this is good enough. As with all other animal interventions, both the 'how' and the 'why' should be carefully considered.

The term 'compassionate conservation' was coined by advocates of an approach which reduces or avoids 'practices that intentionally and unnecessarily harm wildlife individuals, while aligning with critical conservation goals'.[2] The authors of this paper argued that 'moral indifference when causing the suffering of wildlife individuals … risks estranging conservation practice from prevailing, and appropriate, social values'. Some advocates of compassionate conservation assert that the use of any lethal measures is unacceptable, while others argue that while conservationists have a duty to minimise harm, and to use non-lethal measures where feasible, there would be serious implications if all lethal methods were prohibited. Others have it that conservation measures should concentrate on the protection of species or habitats of importance rather than worry about the experience and fate of individual animals.[3] A study by biologists into the 'efficiency' of wire snares to capture red foxes provides a particularly good example of this. They reported: 'We snared 21 red foxes during the 3-year study with only 2 fatal injuries.'[4] While the publication of these findings is welcome, a 10% mortality in any intervention is surely not acceptable. Reports like this can only strengthen the argument for ethical scrutiny of research proposals, irrespective of the objectives of the study.

Perhaps I am being naive, but I can't see why we can't do both conservation and animal welfare. That is, where absolutely necessary, allow for time-limited and humane killing to support a clear objective. Of course, 'where absolutely necessary' is open to debate. Which is why decisions ought not to be the preserve of one person or a remote, unaccountable organisation. I will return to this issue in Chapter 12.

Despite this unresolved debate there are instances where British conservation NGOs seek to hold themselves to account and open themselves up to independent scrutiny. One of the UK's largest conservation NGOs, the Royal Society for the Protection of Birds (RSPB) publishes details of the vertebrate animals killed on its reserves, along with the justification for why control was undertaken. According to the RSPB:

> Lethal vertebrate control on RSPB reserves is only considered where the following four criteria are met: that the seriousness of the problem has been established; that non-lethal measures have been assessed and found not to be practicable; that killing is an effective way of addressing the problem; that killing will not have an adverse impact on the conservation status of the target or other non-target species.[5]

This, along with the statement that killing will only be authorised when certain criteria have been met, is an example of a responsible organisation recognising that this is an ethical issue, regardless of the motive, and that, given the interests and values of its membership, transparency is helpful and necessary. That this is backed by an Ethical Advisory Committee is only to the good. A similar approach has been adopted by the National Trust.

Statutory conservation bodies are sometimes criticised for not considering animal welfare as part of their decision making. There is some evidence for that. For example, in response to the impact on improved pastures of two species of migratory geese that spend the winter on the isle of Islay, the Greenland white-fronted goose and the barnacle goose, a long-term Sustainable Goose Management Strategy was implemented.[6] Although credited to a partnership between various organisations (but including neither the RSPB nor the Wildfowl and Wetlands Trust (WWT)), the strategy was devised and published by staff from the Scottish government's statutory conservation agency, Scottish Natural Heritage (SNH, now NatureScot). The strategy is intended to address the effect of a growing wintering population of barnacle geese which graze on improved pastureland and, it is believed, cause substantial economic loss. As well as measures such as financial compensation for affected businesses, diversionary feeding and crop management, the strategy details a technique known as 'lethal scaring'. This involves shooting at flocks of barnacle geese to kill some and, at the same time, scare away others. Despite the strategy being well researched and well evidenced, the terms 'humane' and 'welfare' do not appear anywhere in the document.

A few months after the launch of the strategy, the RSPB and the WWT submitted, jointly, a complaint to the European Commission about the killing of barnacle geese on Islay.[7] The basis of the complaint was that 'the approach taken in the strategy is insufficiently evidenced to ensure compliance with the Birds Directive'. Like the Islay Sustainable Goose Management Strategy, this is a well-researched and well-evidenced document, albeit one which takes a diametrically opposed view. However, again, the terms 'humane' and 'welfare' do not occur in the document.

Whatever the rights and wrongs of killing geese on Islay, one might think it odd that animal welfare appears to play no part in the deliberations, particularly as there are numerous reports of geese being shot but not killed, and left 'crippled'.[8] None of the organisations involved are animal welfare organisations, and it could be argued that SNH, RSPB and WWT ought to stick to conservation. I don't believe this is an acceptable position. Any organisation involved in sanctioning the killing of wildlife needs to include as an integral part of their strategy development a consideration of animal welfare in their proposals, including an evaluation of the humaneness of available methods. The chosen method should be included in the strategy document as part of a commitment to transparency.

Ringing, tagging and other marking techniques for birds and mammals

Birds, mammals and other animals are individually marked or tagged using a variety of different techniques and for a number of reasons. For instance, 'capture, mark and recapture' techniques can be used to study populations of small mammals and other species that are hard to observe directly. Generally, because the animals do not need to be individually marked and the study is completed over a short period, ephemeral marks such as fur clipping or paint marking can be used. For animals which are to be studied for much longer periods, a permanent, or at least long-lasting, mark is required. For birds, this includes metal and plastic leg rings, wing tags, neck collars and tags applied to the bills of ducks. Larger plastic tags and rings carry numbers which can be read at a distance. Increasingly, as the technology develops, tags and rings carry radio and GPS transmitters which allow for remote tracking over weeks, months or even years.

The most widely used method for birds involves a numbered metal ring fitted to the leg. These rings help in the collection of data, which, accumulated over many years, generate information on patterns of migration, population dynamics and mortality. Birds that are retrapped or found dead provide data on movements and time elapsed since they were ringed, and useful information can also be gathered from the bird at the time of ringing. Fledged birds are ringed after being caught in a net or trap, while nestlings can be ringed when they are still in the nest. There are many methods of capturing birds, the most common being the mist net, although this is most often used for relatively small birds. For larger birds and for birds where mist nets are not suitable or effective there are a variety of other types of non-lethal nets and traps. The British Trust for Ornithology (BTO) oversees and authorises the British and Irish ringing scheme, on behalf of the statutory conservation authorities.

Catching animals for any purpose, even for the most benign intervention and even for a worthy purpose, has a cost. But do we know the cost of the capture and the subsequent handling of birds for ringing? Studies into the impact of capture and handling of wild birds have concluded that bird ringing is not 'cost-free' for the birds concerned, and that researchers need to evaluate and minimise these impacts by adjusting their methodologies, rather than assuming that such impacts do not exist.[9]

Everyone involved must undergo rigorous training under supervision before being granted a ringing licence by the BTO. Much of the training is concentrated around handling and takes account of the different skills needed for, say, handling a crane chick or removing a chiffchaff from a mist net. The equipment is designed carefully for the job – each species has a designated ring type, for example. Record keeping is meticulous and allows for detailed retrospective examination and the development of improved practice and equipment when necessary. For instance, an analysis of data on over 1.5 million passerines caught in mist nets in Britain and Ireland showed that the average direct mortality is just over 1 in 1,000 birds, with the majority caused by being predated while caught in the net.[10] Mortality associated with handling is very low. The analysis ended with recommendations which included using only the most experienced ringers to handle higher-risk species, dealing with the most vulnerable species first, and more efficient net-checking protocols.

Obviously, this does not apply to the handling of cranes, since they are not caught in mist nets but, I was relieved to hear, the mortality rate of ringed crane chicks is not measurably higher than those in comparable circumstances and of a similar age that have not been caught and ringed.

As well as continually keeping ringing practice under review, the equipment needs to be continuously appraised to ensure that injuries associated with rings are minimised. One study, concentrating on three diverse species – brown thornbills, Siberian jays and purple-crowned fairy-wrens – identified a number of injuries associated with rings. These were rectified by adopting new equipment and changing techniques.[11] The Eurasian crane is a long-lived species – up to 20 years and more. Evidence from birds retrapped or found dead showed that older birds were developing abrasions at the top of the tarsal joint caused by the lowest of their colour-coded plastic rings. A decision was made at the 9th European Crane Conference held in France in 2018 to adopt a new design which is believed to reduce abrasion. The impact of these new rings will be monitored.

The development of miniature and affordable radio, satellite and GPS tags has revolutionised the tracking of birds and is transforming the study of migration. As with rings, it is important that the technology does not compromise animal welfare. To avoid bias in the data, it is also important that there is no appreciable difference in behaviour, longevity, breeding success, etc between a tagged bird and untagged one. There is no doubt that, in the early days, when the technology was younger and the weight of the tags was greater, the tags were responsible for some injuries and mortality.[12]

In Britain, all tagging projects other than standard ringing require approval from the BTO's Special Methods Panel, which rigorously scrutinises all such proposals on behalf of the UK statutory conservation agencies. The scientific literature on the effects of various types of tag is large. In a similar fashion to data collected from ringing, studies help to the refine techniques. For example, post-mortem studies of red kites carrying harness-mounted radio transmitters showed that four had sustained serious injuries associated with the harness,[13] underlining the need to monitor their use and to constantly refine the design. Studies can be used to determine whether the effects of tags vary between similar species: GPS devices attached using a crossover wing harness attached to two species of seabird, the lesser black-backed gull and great skua, showed no effect on breeding productivity for either species but for the great skua there was strong evidence of reduced overwinter return rates.[14]

Studies can also be used to establish longer-term effects. There appear to be no examples of this type from Britain, but two from overseas are particularly interesting. A study involving black kites showed no detectable difference between tagged (carried in 'back-pack' harness) and control individuals in key vital rates such as survival probability, recruitment, age of first breeding, reproductive performance and timing of breeding.[15] Other studies have been used to demonstrate the effect of different types of tag on the same species. For instance, a study determined that wing tags have a severe impact on the flight performance of Cape vultures, whereas those attached to the leg appear to have little impact.[16]

Other species of animals are also tagged for the purposes of research. Sophisticated techniques and equipment are used in mammals, and by virtue of the fact that most mammals do not fly and are often heavier than birds, there tend to be fewer problems with either the weight or the bulk of the tags interfering with behaviour. The tagging of mammals is not always benign. A study into the breeding behaviour of the water vole, a species in steep decline in the UK, showed that the number of females reaching adulthood in some of the study sites had dropped by half. This happened only in those areas where the voles had been trapped and released carrying a radio-tracking collar. This phenomenon, where females give birth to (or raise) mainly males, was attributed to raised circulating corticosteroids - a sign associated with chronic stress. In other words, the stress of carrying the collar was altering the breeding biology of the water voles - the very subject of the study. It is to the credit of the researchers that radio tracking was stopped immediately and that the findings were published.[17]

Bats, for obvious reasons, are quite different and need special care. Most species are small, delicate animals and the wing membrane, where metal identification tags are attached, is particularly vulnerable to damage. Unlike bird ringing, there has been sufficient evidence of the injuries that can be sustained from various species of bat that the authorities have introduced moratoria on the use of bands.[18] Strict licence conditions including being able to demonstrate relevant experience apply to any interventions concerning bats, including the fitting of metal bands to the wings. There are fewer concerns about long-term effects of the use of miniature electronic tags since these are intended to be temporary and are simply glued to the skin of the bat's back, but they are still the subject of strict licence conditions.[19]

One has to conclude that, as with any intervention, there is a need to balance the benefits of data acquired by tracking animals against the adverse consequences on the animals' health. Consistent and widespread monitoring of the effects of all types of rings and tags is necessary, as is a consideration of the continued need to mark animals. Again, as well as asking 'how?', we need to ask 'why?'

Introductions and translocations

The UK has a severely depleted fauna (and flora). A measure known as the Biodiversity Intactness Index, which assesses how intact a country's biodiversity is, suggests that the UK has lost significantly more nature over the long term than the global average. The UK is currently placed in the lowest 12% of countries and territories for biodiversity intactness.[20] Considerable efforts are being made to reverse this decline, and while habitat regeneration and rewilding will increase both diversity and abundance, a different approach is needed for species which have been extirpated, locally or nationally, and which are unlikely to re-establish themselves without human intervention. Hence there is increasing interest in animal translocations and reintroductions. An example of a reintroduction is the Eurasian beaver.

While the primary objective of translocation is to increase biodiversity, like any other conservation activity, this must be done while taking into account the welfare of the introduced species and the impact on the animals involved.

Risks to the reintroduced species include the threat of starvation, adverse environmental conditions and disease, as well as the stress associated with capture, transportation and the establishment of a new social structure. Resident species in the release area may be indirectly affected via predation, the introduction of disease and competition for resources.[21]

Although there are detailed guidelines for reintroductions published by the International Union for the Conservation of Nature (IUCN), these are mainly concerned with the preparations for the reintroduction.[22] The IUCN guidelines recommend post-release monitoring to assess whether there are 'unacceptably high levels of disease/adverse welfare/mortality which will impact on the success of the translocated population, or which may present a threat to neighbouring populations'. However, the guidelines acknowledge that if recapture is needed for this purpose, it may only exacerbate underlying problems.

For example, when considering the reintroduction of beavers the main welfare concerns include the beaver's indirect impact on their habitat, thereby affecting the welfare of other animals in the locality, the risk of introducing disease, the methods of beaver capture (from the donor site), health screening in quarantine, transportation, habituation (to the new site), population management (including the avoidance of indiscriminate killing) and an exit strategy, including methods of removal if the reintroduction is deemed unsuccessful or population growth exceeds the carrying capacity of the new locality.

A good example of a relocation project that succeeded in extending the range of a threatened animal was the Cornwall cirl bunting reintroduction project.[23] The cirl bunting in England has for many years been confined to a small area of south Devon. An RSPB-led project sought to extend the range. Although carefully planned from the outset, it was not without its problems. Captive breeding was a failure, prompting a change of approach. Nestlings were collected from the wild in Devon to be reared in captivity near the release site in Cornwall. This worked well, ably assisted by the Zoological Society of London, whose expertise in the hand-rearing of birds was invaluable. There is now a well-established population in Cornwall derived from these captive-reared birds. The success of the project owes much to the attention to animal welfare – successful hand-rearing, release and survival not only ensures a viable population but is indicative of a commitment to the welfare of the individual.

Little evidence exists about the welfare aspects of translocations and introductions, either for the translocated species or for the resident species that may be affected directly or indirectly by the newcomers. Research into the possible impacts on individuals may be warranted, rather than simply relying on the establishment of a new population as a measure of success (but see below).

Predator control to protect threatened species

Animals prey on each other. For many species, catching, killing and eating other species is the primary means of gaining sustenance. Mammals and birds are frequently caught by predators, as adults and especially as dependent young when they are at their most vulnerable.

The fortunes of a number of species of birds of prey has improved over the last 40–50 years. This is primarily because of the introduction of controls over the use of certain toxic pesticides and better legal protection. The recovery of the buzzard has been extraordinary; from being confined to Wales, the West Country and parts of Scotland, this impressive bird now breeds in all counties of England and numbers are higher than ever.

There have been calls for the protection of buzzards to be relaxed because 'there are too many'.[24] I bridle when I hear statements like this. It is meaningless without qualification. My response is 'Too many buzzards … for what?' Given that the generally accepted reason for the recovery of the buzzard is better legal protection (and perhaps because of a more enlightened populous), there's a particular irony in calling for the reintroduction of persecution. Although there is no general relaxation of protection, Natural England (NE) has issued licences to kill buzzards that have been preying on pheasant poults. These licences were issued following a successful Judicial Review challenging NE's approach. As of 2019, nine licences had been issued to kill buzzards to prevent predation of young pheasants, permitting a maximum of 60 buzzards to be killed.[25] There is something deeply unsettling about issuing licences to kill native birds so as to protect a non-native species, the numbers of which have to be annually replenished by captive-bred birds. When one takes into account that a surplus of pheasants means that many carcasses are simply dumped in hedgerows, the killing of buzzards to protect these unwanted birds is simply unacceptable. Although the licences appear to be time-limited (for some unexplained reason the duration of the published licences is redacted by NE) and there is a requirement to use deterrents, neither the buzzard killers nor the responsible statutory agency appears to be under an obligation to develop an exit strategy.

Likewise the peregrine falcon. When I started birdwatching in the 1970s, a sighting of a peregrine was something to get excited about. As with the buzzard, its population and range have recovered to such an extent that many towns now have a pair or two nesting on church towers or other high buildings. Which is surely a cause for celebration.

Some people believe that the recovering populations of various species of bird of prey are causing a decline in songbirds. The organisation SongBird Survival (SBS) maintains that declines in songbird numbers are due, in part, to an increase in the numbers of birds of prey. SBS believes 'in management for balance' and that 'predators … and non-native species need to be carefully managed to help boost songbird numbers'.[26] But this argument falls at the first hurdle since there is no good evidence that the increasing populations of birds of prey affect overall songbird numbers.

It is true that sparrowhawks visit gardens more often than in the past. This is because there are more of them. It may also be because more and more people feed birds in gardens. Like the buzzard, the sparrowhawk has recovered from a population low in the 1960s (largely through the withdrawal of organochlorine pesticides).[27] There is no evidence that the growth in sparrowhawk numbers is affecting songbird populations. Quite the opposite in fact: a 30-year study of songbirds covered the periods of sparrowhawk decline and recovery and found no convincing evidence of sparrowhawk effects.[28] A study of the effects of sparrowhawks on the great tit population in a national park in the Netherlands concluded that despite sparrowhawks taking up to a third of all recently fledged tits during summer, fluctuations in tit populations were unrelated to changes in sparrowhawk numbers.[29] In reality, songbirds produce large numbers of young, only a few of which need to survive in order to maintain existing population levels. Sparrowhawks and other birds of prey are taking songbirds from a surplus that are destined to die anyway from other causes, such as starvation, if not predated.

It is a different story for some ground-nesting species. Although habitat loss has contributed to widespread population declines, numerous studies show that the predation of nests and young is the main reason for breeding failure in ground-nesting waders.[30] Some ground-nesting species, such as the grey partridge, are particularly prone to predation by foxes. Systematic killing of predators has been shown to be effective in increasing successful breeding and fledging of the grey partridge.[31]

Lethal control of predators should not be seen as a substitute for the provision of good habitat and food supplies, and in many cases recovery of prey populations can be achieved through habitat-mediated measures alone. However, particularly for some ground-nesting birds, predator control can provide a valuable additional tool for conservation managers. The decline of the Eurasian curlew is a case in point, with surveys of lowland wet grassland showing losses of almost 39% between 1982 and 2002 and further decline since then.[32] While habitat loss is a major cause, predation by corvids and foxes is also important,[33] and there is some evidence that the rate of predation has increased, perhaps associated with an increase in predator numbers.

While I am not opposed to the licensed carefully planned, *geographically and seasonally limited* killing of certain birds and mammals to protect threatened species such as the curlew, it should only be done once other options have been exhausted, such as scaring and exclusion. The RSPB Curlew Recovery Plan is trialling a series of interventions which will include everything from fencing off nests to predator control.[34]

Of course, any killing must be demonstrably humane. This means competent operatives using the correct equipment under appropriate, independent scrutiny. Finally, for ethical reasons and to ensure killing birds and foxes does not become part of an annual routine, there must be an exit strategy. It is hard to accept that the annual routine killing of predatory animals, whatever the objective, can be described as conservation. It is no more than a sophisticated form of gamekeeping.

The same argument can be made for the use of predator-proof fences. As a temporary measure while alternative measures are developed and deployed, these are justifiable providing they can be used humanely. For example, the use of electric fencing to exclude badgers and foxes from the Greylake RSPB reserve in Somerset has increased the numbers of breeding waders while in other potentially suitable habitats in the vicinity, numbers continue to decline. The underlying cause of the decline is almost certainly the drainage of the moors, but predatory mammals and birds also play a part. Of course, the argument is that this is an attempt to increase the breeding population of waders which would otherwise be declining. I do not disagree with that as an objective. However, the long-term use of predator-proof fences is a tactic not a solution, any more than is the routine annual shooting of predatory species to meet the same objective.

Producing a high-quality environment for a particular species or group of species tends to attract predatory animals drawn in by abundant prey. Measures to reduce their impact can lead to an arms race between the conservation body and the predators. Everything from shooting and trapping to electric fences and widened ditches are deployed in order to protect the hapless birds. Such tactics may be necessary on some small, isolated reserves where rare species need to be protected, but in the long term larger areas of habitat set aside for wildlife, where the impacts of predators are diluted, are surely the solution.

In his book *Rebirding*, Benedict Macdonald argues that a larger, more dynamic landscape with a mosaic of habitat types is the key to increasing breeding success, particularly as it prevents these circumscribed and uniform small areas of habitat becoming magnets for predators.[35] While conservation (and rewilding) is stymied by the lack of the sufficiently large areas of land that are necessary to achieve this dynamism, it is hardly surprising that we have to rely on a network of small, intensively managed reserves. Good examples of this type of reserve are Minsmere in Suffolk and Ham Wall in Somerset, both owned and run by the RSPB.

My long-term (and long-suffering) birding and travelling companion Stuart Reeves describes such places as 'roofless aviaries'. These flagship reserves contain an abundance and diversity of birds which is unrivalled anywhere else in Britain. But, relative to the rest of the country, they are tiny. The breeding birds are attracted into an area which acts as a magnet for predators. As a consequence, a substantial proportion of Minsmere is now protected from the ingress of foxes and badgers by a permanent electric fence. This may have facilitated the successful breeding of avocets, but is the exclusion of ground predators a long-term solution or an acceptable way to treat native wildlife? I am not convinced on either ground. In short, finding an alternative to killing or constraining the movement of our native species of predatory animals by conservation organisations requires bigger reserves.

Predator control on islands

There is a special type of predator control that deserves to be considered in detail. Many seabirds nest in places that help protect them and their vulnerable chicks from predation – such as on cliffs, in burrows and sometimes in trees. Some species nest on small, sometimes remote, islands, to reduce the risk still further. The absence of ground predators in these places means that the birds can nest in relative safety, although aerial predators may still be present.

These islands generally have not been naturally colonised by rodents and larger mammals of the type that raid nests to eat eggs and chicks. However, human visitors over the last few hundred years have led to rats, mice, feral cats and other predators becoming well established on many formerly predator-free islands. Their presence threatens the very survival of certain species, some of which have a breeding range limited to just one or two islands.

Most species have co-evolved with predators and exhibit behaviours and nesting strategies that mitigate their impacts. This does not mean that predators will never be successful but that, subject to the dynamism in biological systems, recruitment to the adult population will be sufficient to maintain the species. For species that have not evolved strategies to cope with a particular predator, such as those on remote islands, the presence of a non-native predatory mammal can mean that breeding success is severely constrained.

Non-native rodents have been eliminated from a number of islands, often with huge resulting benefits for breeding seabirds. As experience grows, more islands are being identified for treatment.[36] For example, on Gough Island, a remote island in the South Atlantic Ocean, the populations of several seabirds including albatrosses and shearwaters, as well as a few species of passerine, are threatened by the introduced house mouse which is preying on both eggs and chicks – and, as recently discovered, brooding adults. Through natural selection the mice have become twice the size of the normal strain; they feast on albatross and other seabird chicks, none of which appear to have developed the ability to avoid predation of this type. Even the adult birds remain stoically at the nest, seemingly unable to react, while the mice eat them alive.

A multi-agency team led by the RSPB has recently completed dropping from a helicopter sufficient anticoagulant rodenticide poison baits to kill all the house mice on Gough Island.[37] These substances are described in Chapter 6, where their use in controlling rodents in the home and food premises is examined. It is beyond dispute that they are inhumane. And that brings us to the central dilemma: is it ethical to use a poison accepted as inhumane to protect the populations of birds nesting on a remote island? In other words, can the suffering of perhaps tens of thousands of mice ever be justified? Ethical dilemmas don't have easy solutions.

For me, it is a matter of utility. Either we kill the mice and protect the birds, or prevaricate while we await the development of more humane substances or control techniques – and meanwhile the population of the birds continues to fall and the chicks continue to get eaten alive.

On balance, and reluctantly, even with the current inhumane methods at our disposal, I support this type of intervention. There are three reasons for this. First, there is an exit strategy which includes the means of preventing new incursions of rodents.[38] Eventually, there will be no more mice and baiting can stop. Second, the suffering of the mice can be 'offset' against the reduction and eventual elimination of suffering of the birds caused by the mice. Third, killing the mice helps to conserve species of birds that would otherwise be at risk.

It is useful to compare and contrast the killing of rodents on Gough Island with other types of killing where species conservation is the main objective. There are two good examples. First, the live-trapping and killing of carrion crows on a commercial pheasant shoot. Killing rodents on a seabird island is inhumane but it has an exit strategy. That is, eventually, it will stop. Live-trapping of crows is also inhumane. The main differences, however, are that this is to protect a non-native species which cannot be maintained without annual releases, and that there appears to be no exit strategy. The killing of crows will not stop so long as the shoot continues. I support the use of rodenticides to kill mice on isolated islands because, in contrast to the commercial shoot, it does have a clear and worthy objective. In the longer term, given the numbers of animals involved, in parallel with the development of more humane methods of managing rodents in domestic environments, alternative more humane methods of eliminating invasive rodents from seabird islands must be developed.

The second example is the killing and ultimate elimination of non-native species which threaten native species from much larger territories. Britain has a number of well-established species of non-native mammals and birds, accidentally and deliberately introduced. These species are often but not always benign. Exceptions include rats, mice, squirrels, coypu, mink and Canada geese – all of which cause problems (or, in the case of the now-extirpated coypu, did cause problems) for wildlife, the environment, businesses and the general public. However, sometimes non-native species threaten the very survival of other species. The ruddy duck is a North American species introduced to England in the 1940s. It freely hybridises with a closely related bird, the globally-threatened white-headed duck.

Although the white-headed duck is not present in Britain, the expanding range of the ruddy duck in Europe threatened its long-term survival in southern Europe. In collaboration with a number of other European countries, in 2003 the UK government took the decision to eradicate the ruddy duck. The intervention has been largely successful. Ruddy ducks have been all but eliminated from the UK, the Spanish and Turkish populations of the white-headed duck are recovering, and hybrids are now rarely seen. Skilled marksmen were licensed for the role, and where it was considered necessary the use of sound-moderated small-bore rifles meant that disturbance to other breeding birds was minimised.[39] This is a good example of targeted intervention that has obvious conservation benefits but where the potential impacts on welfare were minimised by the adoption of appropriate measures.

Concluding remarks

Conservation is not simply about maximising the numbers of species of flora and fauna in all circumstances. Not everything can be conserved, and not everything requires help from conservationists. Priorities have to be set and decisions have to be made. I have some sympathy with conservationists, particularly those in Britain where the size of the available reserves provides few options. Decisions become more and more difficult the smaller and more isolated the areas that are set aside for wildlife.

Where decisions by conservationists involve either the killing of animals or the use of barriers which hinder free movement, then a consideration of both the 'why' and the 'how' is needed. In other words, both the welfare of the animals involved and the ethics of the intervention must be central to the process. All alternative non-lethal interventions must have been considered, and where animals are to be killed it must be done humanely. Competent people using equipment appropriate for the task are essential, as is a degree of third-party scrutiny. Proposals should be subjected to ethical review with input independent of the organisation involved. An exit strategy should be built into every decision or, where this is impractical, there should be a regular review. In particular the annual killing of one species in an attempt to improve the fortunes of another needs careful scrutiny and robust challenge.

Similar principles apply to the collection of data. While the technology advances and offers the prospect of revealing more of the mysteries of migration and population dynamics of birds and mammals, it is important that rings, tags and any other intervention are not likely to cause harm. In other words, the utility of the proposal is an important consideration, and if there is a risk that a substantial number of animals will be harmed, then it must be rejected. Small-scale trials are necessary to demonstrate safety. And every effort should be made to avoid harms and withdraw technology if it is shown to be unsafe. A process of data collection and review must be in place. It has been argued that fitting satellite tags to a small number of a particular species of bird is capable of producing more useful data than applying hundreds of rings to birds of the same species. As this technology becomes more widely available, and suitable for smaller and smaller birds, there has to be a debate about the routine netting of birds for ringing when less risky alternatives may yield more data.

Here on our island, there is little prospect of formerly native mammals re-establishing themselves, and the process of recolonisation by formerly native birds can be painfully slow. Translocations and reintroductions provide the opportunity to improve diversity and perhaps re-establish the trophic cascades lost when many of our native mammals were extirpated. But, as with any other intervention carried out in the name of conservation, it is important that the process is managed in a way that protects the animals' welfare before, during and after release.

It is encouraging that conservation bodies are taking ethics and welfare seriously and have begun to set up internal groups to consider the evidence and the ethical basis of wildlife management decisions. These groups have often independent, expert input which provides external challenge and helps to avoid groupspeak. Conservation bodies have the opportunity to be exemplars of good practice. It is to be hoped that by occupying the ethical high ground they will, eventually, influence other landowners and land managers to adopt a more enlightened attitude.

There is no point in being coy about this. Statutory and NGO conservation organisations who make difficult decisions should have the courage of their convictions and be open about the process and the decisions reached. I believe that, particularly with membership organisations, the more transparent the decision making, the better. Assume that your members will want to know. You will be rewarded by the loyalty of better-informed and thoughtful members.

8

Recreation, Sport and a Little Food

The use of animals in sport and for recreation has a long history. The hunting and killing of animals goes back, surely, to the earliest humans. Yet, while hunting animals is clearly important in hunter-gatherer societies as a source of protein and other nutrients, recent studies suggest that some types of hunting have always served other purposes:

> Men, however, hunt large and unpredictable game as a political strategy aimed at maximizing their social currency by participating in competitive masculine displays ... Hunting is therefore seen as a form of costly signalling whereby males 'show off' in order to gain status, attract women for pair bonding, and to attract other men to form alliance partnerships.[1]

The mischief makers among us might argue that much of today's hunting, shooting and angling serves a similar purpose – particularly because, for a variety of reasons, in many cases the quarry is never intended to be eaten. For instance, the vast majority of line-caught coarse fish are put back in the river alive, and non-trivial amounts of shot pheasants are either dumped or buried. In reality, of course, the recreational and social aspects of hunting and the harvesting of wildlife for food are closely linked – and no doubt that was also the case for our earliest ancestors. But many of those who go shooting and fishing are motivated primarily by the experience of the outdoors, the social aspects and the thrill of 'the chase'. Returning home with a brace of pheasants or grouse for the pot is, for many shooters, much less important than the taking part. Similarly, many anglers, even on a successful day, bring no fish home.

But let's not dwell on that, because a substantial amount of hunting, shooting and angling does end up with something for the plate. What little sea fishing I do ends, disappointingly infrequently, with me grilling mackerel fillets on a barbecue. Many shooters and anglers do the same thing but with greater success.

We must not dismiss the professional: the deer stalker working to manage red deer numbers in Scotland or patrolling woodland in England and stalking deer to limit the impact of several species on vegetation. In both cases, the venison will likely end up on someone's plate. And at least a small proportion of Britain's favourite 'game' bird, the ring-necked pheasant, released in the tens of millions, ends up in supermarkets and butcher's shops.

It is, I hope, obvious that obtaining food from the environment does not always require the land to be farmed or otherwise managed. The 'harvesting' of birds and fish may draw on wildlife resources where human intervention is absent or at least minimal.[2] In reality, as we will see, this is rarely the case here in the UK.

Where the meat ends up is of secondary interest to us from a welfare perspective. However, I am reminded of the words of the TV chef and food campaigner Hugh Fearnley-Whittingstall: 'This animal died so you could eat it. The least you could do is eat it all.' It's a sentiment I can agree with. Our primary concern, though, is the welfare of the hunted animal and the fate and experience of other animals caught up in the process (including other species that are killed because they are deemed inconvenient to shooting, fishing and hunting interests – see Chapter 6). You may ask yourself, does the venison stew or roast pheasant I am about to eat derive from animals whose welfare has been taken into account? And, importantly, how do I find out? We will delve into that shortly.

The racing industry is responsible for another widespread and popular use of animals for recreational purposes, and although animals are not deliberately killed or injured in either horse or greyhound racing, there are multiple concerns about the welfare of the animals involved. Both types of racing are covered in the last part of this chapter.

Hunting with dogs

I was never a fan of hunting mammals with dogs and was pleased when it was banned by the Hunting Act 2004, imperfect though that legislation is.[3] The Act banned the hunting of deer, foxes and other mammals with dogs.

A debate about hunting with dogs had run for decades. Proponents argued that it was essential for 'pest' control, and that it was an important way of maintaining countryside traditions. Opponents said it was an excuse for cruelty and that, if 'pest' control was necessary, there were humane alternatives. In the event, the Act landed in the statute book but the debate rages on, with most hunts continuing with some activities using dogs to this day.

I try not to get mired in the detail of the law. There are plenty of other people more knowledgeable than me and all too happy to do that. But if the original intention of the Hunting Act was to prevent the deliberate act of setting one animal on another to initiate a violent death, it appears to have failed. As I explained in Chapter 2, it's not a quick death; both the chase and the kill are undoubtedly inhumane. While, on the face of it, the welfare of the hunted fox, deer and hare in Britain substantially improved with the introduction of various prohibitions on the hunting and killing of these species with dogs, the illegal killing of wild mammals with dogs continues. Many hunting groups with packs of dogs continue to roam the countryside, ostensibly following artificial scent trails or exploiting other loopholes in the legislation. I find it difficult to accept that the frequency with which a pack of dogs 'accidentally' comes across a fox and 'accidentally' tears it to pieces is simple happenstance. And when one takes into account the way in which foxhounds and horses are treated as disposable tools of the hunt, one is forced to conclude that the entire activity is conducted with no regard for the sentience of the animals they use.

It is disappointing to see the various ways that people have found to hunt using loopholes within the Hunting Act – or by ignoring the law altogether. I can only hope that enforcement effort is increased and where necessary the Act is amended to ensure that loopholes are closed. Recent decisions taken by the National Trust and other landowners to prohibit trail hunting on their land is, one can but hope, the beginning of a trend. And perhaps the beginning of the end.

While formal, organised hare coursing appears to be extinct, illegal coursing is apparently still rife, difficult to police and probably linked to illegal betting and violent crime. Other illegal activities involving the deliberate injury and killing of wild mammals with dogs, such as badger baiting, continue, although this is probably now rare. Similarly, there are occasional reports of organised illegal dog fighting and cock fighting. All of these cause obscene and often fatal injuries to the animals involved. As well as being illegal, they are morally indefensible. I don't know what motivates these people and I don't wish to find out. But I'll state this – the hunting and killing of mammals with dogs, whatever the species, whatever traditions and whatever rituals are involved, is an unacceptable abuse – whether legal or not. It needs to stop.

Killing animals for food and recreation

Birds, mammals and fish are killed both for food and for recreation. There is no material difference, in the moments leading up to death, between the experience of an animal killed for food and one killed for recreation – or

146

indeed one that is exterminated because it is considered to be a 'pest'. I will not draw a distinction in describing the process.

Let's start with birds. The three most commonly taken avian quarry species in Britain are the ring-necked pheasant, the red-legged partridge and the red grouse. While the latter is a native species, the other two are introductions. Tens of millions of pheasant and partridge poults are released annually to maintain numbers for shooting. Producing millions of pheasant poults is a huge business and requires substantial inputs – feedstuffs, disease and parasite control, the use of 'bits' to prevent feather pecking and cannibalism,[4] and, once they are released, the killing of large numbers of predatory species such as foxes, stoats, weasels, crows and sometimes buzzards.

While some pheasant breeding is small-scale and largely protects the welfare of the birds, a substantial proportion is not. A high proportion of the birds released in Britain are imported from continental Europe, notably from France and Poland, either as hatching eggs, day-old chicks or poults. Many of these are produced in large premises where the birds are husbanded in conditions that would be prohibited for domestic hens, although many smaller breeding units use much less intensive methods. Similar arrangements apply to the red-legged partridge, albeit on a smaller scale. In summary, wild shooting it is not.

Unlike pheasants and red-legged partridges, red grouse are not artificially reared. The red grouse is considered to be a subspecies of the willow grouse, a species that occurs widely across the Holarctic. In Britain, it is a bird of the uplands and particularly of heather moorland. It is exploited as quarry through two different approaches to shooting. The first, known as 'walked-up' shooting, involves little prior intervention and is essentially a form of rough shooting with birds taken opportunistically and in small numbers as they are disturbed. The second, driven grouse shooting, is a different thing altogether. This involves beaters driving the grouse towards the shooters, who are lined up across the moorland hidden in shooting butts. This requires a high density of birds to ensure that large numbers are available to be driven over the guns, and the moorland is managed intensively to that end.

Mark Avery's book *Inglorious* lifts the lid on driven grouse shooting, and there is no need to repeat his compelling arguments here.[5] Suffice to say, as with pheasant and partridge shooting, the production of an annual 'shootable surplus' of red grouse for driven shooting involves considerable manipulation of the environment. Estates devoted to driven grouse shooting alter the landscape and routinely kill predatory birds and mammals that are deemed to be a threat to the shootable surplus of grouse. That includes both the legal killing of animals such as corvids, stoats and foxes, and the illegal killing of animals that are in theory protected, particularly birds of prey. There is little or no third-party scrutiny of the killing and no information about the impact on populations of the species involved. The illegal killing of hen harriers on grouse moors has brought the species to the brink of extinction in England.

Measures to control parasites and disease, and to improve the fecundity of the grouse – such as the provision of medicated grit in plastic trays that litter the moors – are a far cry from anything remotely like the typical experience of genuine wildlife. Natural it isn't, and although it is rarely described as farming, there are many parallels with the production of livestock.

In terms of numbers, the shooting of these three bird species represents the vast majority of wildlife shot ostensibly for the pot. I say 'ostensibly' because there is good evidence that the majority of pheasants released for shooting are neither shot nor eaten by humans. Roadkill, predation, disease and starvation reduce the numbers available for shooting.[6] In any case, the market for pheasant meat is hardly healthy if the frequency of reports of carcasses being buried or simply dumped is representative.[7]

Another form of sport shooting known as wildfowling, the shooting of wild ducks and geese, involves minimal manipulation of the environment. Relatively few ducks are reared and released for shooting, although the use of flight ponds baited with grain to attract wild ducks is a widespread practice.[8] Small numbers of snipe, golden plover and woodcock are shot for the pot but none of these activities involves significant environmental manipulation.

The manipulation of the environment for deer, the main mammalian quarry of shooting 'for the pot', is less intense. There is no need to kill deer predators, because none remain; the wolf, brown bear and lynx were extirpated centuries ago. Indeed, despite seasonal stalking of red deer in Scotland and continuous efforts to reduce deer numbers in England to limit their impact on forestry, the numbers of all deer species continue to rise. The numbers are now so high that they are affecting woodland maintenance and regeneration, and this has led to repeated calls for more control to reduce numbers.

Shooting of birds and mammals: welfare concerns

There are two broad categories of gun used for shooting birds and mammals in Britain, the rifle and the shotgun. Shooting of birds generally involves a shotgun, and although small-bore rifles have been used for large 'game' birds such as black grouse this is not common. The law requires deer to be shot with a rifle of an appropriate calibre. Rifles and shotguns are also used for shooting foxes and rabbits.

For obvious reasons, the keeping and use of shotguns and rifles is tightly regulated. You need a firearms or shotgun certificate issued by the police to possess, buy or acquire a firearm or shotgun. You must also have a certificate to buy ammunition.

The conditions that apply to gaining a shotgun certificate are different to those for a rifle. In essence, you need to provide a very good reason to get a firearms certificate for a rifle, whereas the onus is on the police to provide grounds for refusing a shotgun certificate. For either, you need to be a fit and proper person, with no history of mental health problems and no criminal record. To obtain a firearms certificate for a rifle you will need to be able to demonstrate, in addition, a 'need' to use the rifle. For example, you may own a piece of land where fox control is carried out, or perhaps you are a professional deer stalker.

You might think that if you wish to shoot birds or mammals, an ability to shoot accurately and to kill the animal 'cleanly', and to identify birds and mammals correctly, would be prerequisites for the granting of a firearms or shotgun certificate. This is the case in many countries. In Italy, for instance, as well as the checks necessary to protect public safety, applicants must complete a comprehensive course that covers shot placement and species identification. This is followed by an examination prior to a licence being granted. In the UK, there are no mandated competence requirements, and despite shooters being encouraged to get appropriate training and follow codes of practice,[9] there are no data about the proportion that do, and there is no widely adopted system of training and re-accreditation.

In Chapter 6 I argued that the IEP's study of badger killing was perhaps the gold standard of published studies in determining the humaneness of killing wild animals. In contrast, there are very few data on the welfare of birds that are shot, whether for recreation, the pot or as 'pest' control. What little is available comes from continental Europe or North America. A review of the welfare of killing (and taking) of birds under the Wildlife and Countryside Act 1981 concluded that evidence about the welfare of birds covered by General Licences issued under the Act is scant.[10] There are no published data on wounding rates or non-retrieval rates, both of which are essential if comparisons between the competence of shooters and the reliability of equipment are to be made.

Shooting birds with a shotgun is not a precise operation. The effectiveness of the shooting depends on a number of factors: not only the competence of the shooter, but also the prevailing weather conditions, the distance from the target bird, the choke on the shotgun barrel and the type of shot involved, etc. The bird may get hit by several pellets, from which it may die instantly or at least quickly. But if it is hit with only a small number of pellets in non-vital organs, it may only be wounded.

The fashion for 'high birds', where it is considered 'good sport' to shoot at pheasants flying high overhead, only increases this risk. Good shooting practice requires that wounded birds are retrieved promptly and dispatched humanely, but reports of the public finding wounded birds in gardens in the days after a shoot suggest that this may not always be done as efficiently as it should be.

In the absence of any independent scrutiny, it is impossible to reach a firm conclusion about the impact of shooting on the welfare of wild birds. But the anecdotal information that does come to light suggests that it is not good.

The shooting of deer is different. Although it is not a legal obligation, the vast majority of deer stalkers and others who shoot deer have formal qualifications which cover, among other things, shot accuracy and animal welfare.[11] One might expect that the efficiency and humaneness of deer shooting would be well documented and understood. However, the evidence is equivocal. The objective when killing a deer (or any other animal) is that it is rendered instantaneously and irreversibly unconscious and hence insensible to pain. It follows that the shooter should aim to place a single shot in the area of the body most likely induce that state. However, there is a balance to be struck. The ideal target area is the brain but the head is too small an area to ensure accuracy at range, and therefore the recommended area in the deer is an area of the thorax (chest) immediately to the rear of the point of the shoulder. An accurate shot placed there will damage the heart and the associated great vessels, that is the major arteries and veins. The concussive effect and the damage to the thorax will cause a catastrophic loss of blood pressure leading to unconsciousness and death.

Studies on the humaneness of shooting deer have relied either on the examination of carcasses sometime after shooting or on reports by shooters and the subsequent examination of the eviscerated carcasses in a game larder. Despite the different ways in which the efficiency of shooting has been recorded in those studies, the results are generally consistent and suggest that between 10% and 14% of deer are shot but not killed outright.[12] However, in contrast to the IEP's report on badger shooting, there are no data on the animals shot at but not retrieved – which might suggest that these figures are an underestimate.

Firearms in the right hands and used for the right purpose are useful tools. I've used them myself – in my youth, I did a little small-bore rifle shooting, and when I lived in Belize I hunted the plain chachalaca, a bird about the size and shape of a pheasant. Later in my career, and all too often, as part of government disease control measures, during the 1990s I had to use a bolt gun to kill cattle suspected to be infected with BSE. I haven't used a firearm for 30 years but I don't think that shooting should be banned. Nonetheless, better controls to ensure that animal welfare is properly taken into account are badly needed.

When it comes to shooting wildlife, I have no doubt that the right weapon in the right hands, used at the correct range, can be an effective and humane method of killing. But like the use of traps and poisons detailed in Chapter 6, there is precious little regulation to ensure that this is the case. People killing farmed animals are legally obliged to undergo training and obtain a licence. These licences may be withdrawn when the law about humane killing of animals is not followed – every year, several slaughtermen lose their licences because they didn't comply with the licence conditions. Similar rules apply to the killing of animals in research establishments.

With the exception of deer shooting, few people in Britain have formal qualifications which attest to their accuracy at shooting wildlife, their skills in quarry identification, or the ability to humanely dispatch wounded birds. When you take into account the evidence about sentience and the efforts expended to ensure that animals are killed humanely in other circumstances, what is it about shooting that provides the free pass? I don't know – but I suspect it might have something to do with the influence of those involved and their ability to head off any attempts to introduce better regulation.

Angling

Angling, in all its forms, is a very popular activity. It can take you to idyllic and beautiful locations where you can immerse yourself in the environment to the extent that it doesn't much matter whether you are successful in catching a fish or not. Of course, it is even better when you catch something.

When I lived in Belize I was lucky enough to go fishing on the barrier reef. There are few more exciting experiences than when a barracuda takes the bait and the reel begins to run. Better still is landing a 10 kg fish a few minutes later.

However, as the evidence mounts about the sentience of fish and the harm that angling does to them, I am becoming less and less comfortable with the activity. Fish are vertebrates and hence have a nervous system that is comparable to that of birds and mammals. The evidence that fish can feel pain (as it is defined by scientific consensus) and will take steps to avoid aversive stimuli is unequivocal.[13]

Although these conclusions were derived from fish in laboratory conditions, there is every reason to assume that the experience of fish in the wild is similar. There are ways of making improvements to the humaneness of shooting and trapping of mammals and birds, but such changes seem less likely when it comes to angling. The increasing use of barbless hooks and the more sympathetic handling of the fish once it has been landed are steps in the right direction. But the experience of being hooked, dragged through the water and handled in an alien environment are integral to the activity of angling, and it seems unlikely that much more can be done to make it more humane. When the express intention is to return most or all of the fish you catch, as it is with coarse fishing in the UK and increasingly with fishing for salmon and trout, one has to ask whether the harm done is justifiable.

Difficult decisions

What does this all mean for the concerned consumer and citizen? There's a lot to take in. I've thought about these issues for some time but I still find decisions about what to eat and what to avoid difficult. For what it's worth, here's my current take on eating wildlife.

I've made the decision not eat to 'game' birds. No single factor was crucial in reaching that decision but, in combination, the inhumane methods of rearing birds for release, the routine and inhumane killing of mustelids, corvids and foxes to protect the released birds (detailed in Chapter 6) and the lack of data about whether the shooting is humane were sufficient. In contrast, I still eat venison. This is because there is no killing of predators (in Britain full-grown deer have no natural predators) and because deer stalkers are almost universally trained to a standard. Of course, improvements can be made but, unlike for 'game' bird shooting, there are at least some data on shooting efficiency and a commitment to good practice through training and certification.

I no longer fish for barracuda, though in recent years I have fished for mackerel and sea bass on Chesil Beach. But increasingly my conscience bothers me. Does the exhilaration of catching a fish override concerns about the welfare of line-caught fish? Perhaps this is why my rods and reels have stayed in the cupboard these last two years. I still eat commercially caught sea fish, but given that these have been caught in a vast net, dragged up from the depths only to suffocate on the deck of a boat, it might be argued that the line-caught mackerel that I landed myself is a better alternative.

The decision you make about whether to participate in shooting and fishing, or to eat the products of either, is yours and yours alone. I only ask that you take note of this chapter and Chapter 6 and consider the welfare of the animals involved before taking decisions about whether to kill and/or eat them.

Animals in sport

Horse racing

Horse racing is big business. The total economic contribution of horse racing to the UK economy is estimated to be in the region of £3.5 billion, with over 85,000 jobs supported by the industry.[14] After football and rugby it is the third most popular spectator sport in the UK, with 1,450 fixtures held annually. If you attend the races, or place bets on the outcome, you ought to be aware of the arguments for and against racing.

Proponents of horse racing argue that these animals, presumably because of their value or potential value, are well cared for and provided with individual attention, a good diet and the best of veterinary care. Despite that care it is the injuries and fatalities that occur during horse racing that cause the most disquiet and controversy.

Are racehorses exploited? There is no doubt that despite or perhaps because of the popularity of horse racing and the money that it draws, horse racing comes at some cost to the horse. Advocates for animal welfare argue that the cost is too great. For example, the League Against Cruel Sports (LACS), an organisation which campaigns for changes in horse racing, among other things, believes that mortality and injury rates are too high, particularly for the longest and most gruelling events. LACS believes that animal welfare could be improved by reducing the number of horses in races, making fences more manageable, and shortening the length of the races.[15]

The use of the whip in horse racing is routine, and although it is tightly controlled by the racing authorities, its use remains controversial. Proponents argue that the whip is merely an encouragement for the horse to perform at its best. LACS, however, believes that use of the whip urges the horse to perform beyond its normal capabilities and can result in injuries and stress.[16] There is good evidence that although horse skin is thicker overall than human skin, the part of the skin that is thicker does not insulate the animal from the pain that is generated by a whip strike, and that humans and horses have the same basic anatomy capable of detecting pain in the skin.[17] Given the evidence that horses do not run faster when whipped,[18] and that modern padded whips intended to cause less injury still cause welts like conventional whips,[19] the argument for their continued use is wearing very thin. Norway has banned the use of the whip except when safety is threatened, and there are types of racing in the UK where the whip is prohibited or severely restricted.[20] We must hope that the trend continues.

There are strict controls over the use of any physical means to drive animals in markets and slaughterhouses, to the extent that it is effectively an offence to strike an animal in a slaughterhouse lairage.[21] It says something about our inconsistent values that the horse, which enjoys an otherwise exalted status in our society, may be struck repeatedly in the name of entertainment and profit, and yet the humble cow is protected from similar physical insults.

The racehorse, like any other high-performing animal, is fed a high-energy, high-protein diet. Unlike a fattening pig or a broiler, this high-quality diet is not to prepare the horse for slaughter for human consumption, but to ensure it has the requisite bone and muscle to race and the energy to maintain a high speed for a substantial period of time. Sufficient amounts of energy and protein cannot be supplied by grass and conserved forage alone, so racehorse diets consist of a high proportion of grains and concentrates. These diets, particularly when combined with training, appear to make racehorses prone to developing painful gastric ulcers to the extent that a very high proportion of racehorses in training are reportedly affected.[22] The exact mechanism is not fully understood, but a low-fibre diet appears to increase amounts of stomach acid, which increases the risk of ulcers developing. Gastric ulcers are associated with a variety of clinical signs in affected horses including increased risk of colic, poor performance, diarrhoea, weight loss, lack of appetite and changes in behaviour. Treatments are available, but the most effective medicine, omeprazole, must be withdrawn from racehorses in training.

Is the prevalence of gastric ulcers in the racehorse and the apparent necessity of treatment with antacids acceptable? Despite the apparent lavish care given to racehorses, this common condition suggests that all is not well with the prevailing conditions and that, despite the value of the animals, some aspects of racehorse husbandry are not so different to the worst of intensive food animal production. Which brings us to race-day injuries and fatalities.

Horse racing is, without a doubt, dangerous. The risk of an injury, sometimes a fatal injury, is high. Some non-fatal injuries such as a foreleg fracture often necessitate slaughter at the track because the prognosis is so poor. Fatalities and injuries resulting in immediate slaughter are understandably distressing to racegoers and the general public. There is pressure to reduce the numbers of injuries. Some people would go further and would like to see horse racing banned.

It is clear that the British Horseracing Authority (BHA) has taken this criticism to heart, as considerable efforts appear to have gone into risk reduction. The BHA has worked with the RSPCA to effect improvements to fence and course design in an effort to improve welfare and reduce race-day fatalities.[23] Figures produced by the BHA do show a reduction over time in fatality rates for all racing. However, the proportion of race-day horse fatalities differs depending on the type of racing. One would assume that the risk of injury and death for horses would be greater where the race involve jumps, and the data bear this out. According to the BHA, fatalities in jump racing show a five-year average of 0.38% of runners (2020). This has fallen from a five-year average of 0.49% in 1998, suggesting that efforts to reduce risk have had some effect. The risk of an equine fatality in flat racing remains substantially lower than for jump racing at between 0.1% and 0.2% of runners.[24]

Regardless of these improvements, the fact remains that horse racing carries substantial risk of injury to the horse (and the rider). It is unlikely that the activity can be made risk free, irrespective of the extent of tweaking of the course, reduction in the height of the fences and length of the race, and changes to the rules about the use of the whip. Greater efforts to reduce risks and further improve animal welfare would reduce the excitement generated by the spectacle, which, in turn, would likely reduce public interest.

Away from race-days risks, there are similar concerns about the 'wastage' of younger horses in training. Various studies show that a substantial proportion of those that never race or simply have to be rested for a while during training suffer from a variety of musculoskeletal problems. For instance, lameness was the most important condition causing horses to miss training, with stress fractures, joint problems and 'sore shins' being the most important causes of lameness. One study showed there had been little change in the preva-lence of these conditions over the previous 20 years, while another concluded that reducing these problems required changes to the training regime for two-year-olds.[25] Note that a horse is not fully physically mature until it is 5–6 years old.

The final consideration is the impact on behaviour. The horse is a social animal and given the opportunity will form mixed-aged and mixed-sex herds. The rearing and training of racehorses (and indeed almost all domestic horses) constrains this natural behaviour, as many are kept isolated for long periods. Breeding is carefully controlled.

I confess to having little interest in horse racing. There is enough concern about many aspects of the sport – from 'wastage' in young horses to the use of the whip, to injuries and fatalities incurred during training and racing – for me to avoid any involvement. On the scale of animal exploitation, it is hardly the worst but neither is it the best. Efforts to improve things seem genuine, and the racing industry does seem to be working towards a growing transparency. But, as with selection for rapid growth rates in broiler chickens, the fate of the racehorse is inextricably linked to the speed expected of it. The welfare of the slower-growing broiler might be protected, as would the welfare of the less pressured racehorse. But in both cases this is exactly the opposite of each sector's objectives. As in the case of the broiler, I believe the most effective measure is to vote with your feet. Fewer spectators and fewer punters might mean that the relevant authorities sit up and take notice.

Greyhound racing

Greyhound racing is often described as the poor relation of horse racing, but it is not a minor sport. It is Britain's fifth most popular spectator sport, with an industry which has a £1.3 billion off-course betting turnover and core industry income of £119 million.[26] However, the internet is full of lurid reports of greyhounds being reared and trained in terrible conditions, suffering poor care and miserable deaths at the hands of uncaring owners. Many of the organisations responsible for these reports call for a ban on greyhound racing because they believe it is inherently inhumane. However, there are few objective reports.

Greyhound racing in the UK was largely unregulated until 2010. New regulations introduced that year now require greyhound racing tracks to be licensed, and to be operated with conditions that include veterinary attendance and facilities, kennels, and for all racing greyhounds to be identified with an ear mark and a microchip and registered on the database set up under the regulations.[27]

A 2015 report by the House of Commons Environment, Food and Rural Affairs Committee (EFRACom) on greyhound welfare noted that the implementation of the 2010 regulations had improved the welfare of greyhounds at race tracks.[28] However, the report voiced concern that data on injuries and euthanasia were scant, and that this meant assessing welfare remained difficult. Kennelling standards away from the track were of concern.

A report by the Dogs Trust reinforced this view by detailing poor conditions at greyhound kennels.[29] The Trust argued that the 2010 regulations focused too much on welfare measures at the track (where greyhounds spend just 10% of their time) and provided no legislative protection for greyhounds during breeding, kennelling, transportation and retirement. These 'out of sight' premises, the Trust argued, where greyhounds spend the majority of their time, will continue to evade scrutiny. The EFRACom report recommended that common welfare standards be developed for all kennels, and that an independent body verify those standards.

All organisations, including EFRACom, are concerned about the fate of retiring greyhounds, particularly those that cannot be rehomed. There are gaps in the data, which suggests that some are euthanised clandestinely. EFRACom believed that high welfare standards require financing and that bookmakers should contribute, given the profits they make from this sport.

In summary, while there appear to have been some welfare improvements, particularly at the race track, the lack of progress in effecting improvements in kennelling, rehoming and addressing injuries is in contrast to the progress made in the horse-racing sector. I am not convinced that the welfare of greyhounds is given enough attention by the organisers, spectators and supporters of the sport. As with so many of the issues that affect animal welfare, legislation can only go so far. Further change has to be driven by the citizen, particularly those with an interest in the activity.

9

Pets: Exploitation Begins at Home

Pets or 'companion animals' are important to a lot of people, and consequently the business of pet care is important. The value of the pet food market is estimated at £2.9 billion and consumers spend another £900 million on accoutrements such as leads, bowls and dressing up their pets.[1] That's because there are a lot of pets. In Britain around 41% of households have at least one companion animal.

The most common pet animals are dogs and cats – by some considerable margin. It is estimated that in 2020 there were 9 million dogs in Britain, with around 23% of households having one or more. When it comes to cats, the estimate is 8 million, with around 16% of households having at least one. The next most popular pets are the rabbit (1 million) and indoor birds (1 million). While the number of households keeping either a dog or a cat has fallen in the last few years, the totals for both have increased, indicating that keeping two or more of either species is becoming more common.[2] However, this trend may have been reversed as pet ownership, particularly of dogs, appears to have increased during the COVID lockdown.[3]

Our increasingly urban existence means many of us rarely experience farmed animals or wildlife close up. For some, experience of close contact with animals comes primarily from the pets in their household. Pets bring companionship and joy to millions of people. I am no exception. I was brought up in a household with a dog, and for the last 35 years I've had one or more cats living in my home. Like many people, I enjoy having cats around. I know that many others feel the same way about dogs, rabbits, fish and even snakes.

Pets are good for you, too. As well as the obvious health benefits from regularly exercising a dog, the mere presence of a pet in a household is considered beneficial to mental health.[4] For example, dogs trigger similar neural pathways to the parent–baby bond, and so help to reduce loneliness and depression. Pets may be of particular value to older people and patients recovering from major illness or with long-term health problems, but on the downside, a reluctance to part with a pet may lead to non-compliance with health advice on, for instance, the avoidance of allergens. Older people may delay seeking medical attention for fear of losing a pet should they be admitted to hospital. Despite that, on balance, pets are a good thing for physical and mental wellbeing.

I can attest to that. In 2010 I suffered a potentially life-threatening illness. Prompt treatment may have saved my life – it certainly helped speed my recovery. At the time, we had two cats at home, and I found their mere physical presence a great help while I recovered. They sneaked into my room and onto the bed in the mornings after my wife left for work, and although I didn't measure what effect they had on my blood pressure or my heart rate they gave me a great deal of comfort. Now both in their early teens, they plod on, leaving tufts of black hair on the rugs and monopolising the comfiest chairs.

The exploitation of dogs and cats

Are my cats and other pets exploited? Many people would say no. The animals may be confined for a large part of the day, and their natural behavioural repertoire is inevitably constrained, but this is offset by the care, food and shelter that is provided by the owner. But is it?

Let's look at the most commonly kept pets: dogs and cats. Whatever the concern you might have about the keeping of pets, you would be hard pressed to argue that the experience of most dogs and most cats is comparable to that of the caged hen of Chapter 5 or the snared fox of Chapter 6. However, we should not forget that exploitation is broadly defined and includes all forms of human intervention, however subtle and apparently benign.

Compare, for example, the life and experience of a pet dog with its closest living relative, the wolf. A wolf is a carnivore which generally lives in a social group. Day-to-day life involves foraging, scavenging, hunting and a complex social life. Bonds between wolves are strong and are constantly reinforced by social interactions including, particularly in younger animals, via play. In contrast, most pet dogs are solitary and are prevented from roaming, thereby inhibiting the full spectrum of normal canine behaviour. Most are prevented from routine socialising with other dogs, and few are allowed to breed.

Of course, while many dogs do have canine companions, a pet dog will bond most closely with its owner. Indeed, that bond is the product of 10,000 years of selective breeding for that very trait. It's extraordinarily strong and can develop within minutes. This rapid development of social bonds reflects the dog's ancestry: keeping up and in with the pack is essential to the wolf's survival. However, that has its own problems – with an estimate of 20–80% of dogs experiencing separation anxiety when left on their own.[5] That is a pretty damning statistic.

Human-mediated deformities

A similar concern is the frankly appalling way that some dog breeders have ruined a number of breeds of dog by selecting for extreme conformation such as extremes of size, flat faces, bulging eyes and other deformities.

Take, for instance, Pugs, Bulldogs and French Bulldogs. These breeds have been selected to have a short skull shape (known as brachycephaly) which gives the appearance of a flattened face. The most severely affected individuals in these breeds will suffer throughout their lives with a variety of conditions such as breathing difficulties, often associated with overheating, sleep apnoea, regurgitation, eye disease, and an inability to mate or give birth naturally.[6]

And all because they look 'cute'. It wasn't always so. There are numerous websites comparing the physical appearance of typical examples of certain breeds today with those of the same breed 100 years ago. The more extreme conformation and disfigurements are recent 'improvements'.[7]

Similar problems exist at the extremes of size. Giant dog breeds are susceptible to growth-related problems thought to be linked to rapid long-bone growth, a high prevalence of bone cancer and heart disease, and many are prone to developing potentially fatal gastric torsion after eating.

At the other end of the spectrum so-called teacup varieties of toy dogs suffer from numerous health conditions related to size reduction, including bone fragility and limited bone growth, particularly in the skull, where development can stop before the bones fuse together (this is so common that the soft spots have become an acceptable part of some breed standards in the American Kennel Club).[8]

These are just a few examples of the problems that beset pedigree dogs. Most breeds are associated with one or more inherited conditions, particularly where inbreeding to select for desirable characteristics is practised. Some of these problems can be managed with medical and surgical interventions but some are chronic and untreatable, leaving the animal to experience a lifetime of suffering. It would be far better to avoid them altogether.

We could and should be better than this. Here we have an area where many of us could make a direct and immediate difference to animal welfare. In spite of all of the evidence and the blindingly obvious deformities that bedevil these unfortunate creatures, brachycephalic, giant and tiny breeds remain stubbornly popular despite the best efforts of veterinary organisations.

We could choose not to buy these wretched animals – and yet we do. Why? We seem to have an ingrained proclivity for pets (cats and rabbits can be similarly afflicted) with extreme conformation in spite of the clear evidence of the harm that these undesirable inherited deformities cause. There's a handy way to avoid these problems. Cross-breed and mongrel dogs are far less likely to be affected with inherited conditions. Consult a vet before you take the plunge, and select a 'dog-shaped dog' of medium size, and you are unlikely to go far wrong. It will be cheaper too. The cost of surgery to correct certain deformities and the costs of long-term treatments can be eye-watering. And don't assume that pet insurance will cover it. The premiums for certain breeds – those with the worst record for deformities – are similarly eye-watering.

We may rail about the conditions that farmed animals and research animals are expected to endure (often without having had any direct experience). But, to be clear, there are no other animals kept for any purpose which are deliberately selected for a trait perceived to be endearing but which will also cause lifelong suffering. It is not as if it is difficult to resolve. Three generations of outbreeding would almost certainly eliminate these inherited defects. But arrogance, intransigence and vanity prevail, fuelling the demand for these manufactured freaks.

Perhaps recent legislation will assist in achieving the objective of better conformation. Regulations in effect from 2018 state:

> No dog may be kept for breeding if it can reasonably be expected, on the basis of its genotype, phenotype or state of health that breeding from it could have a detrimental effect on its health or welfare or the health or welfare of its offspring.[9]

One would have thought that would apply to a substantial proportion of the specimens currently waddling about – but, to date, there have been no prosecutions. If and when there is, it will no doubt be watched carefully by many of the dog breeders in this country.

I'm biased of course, but I believe that the cat is the perfect companion animal. In contrast to dogs and most other pets, the amount of effort required to keep a cat is minimal; they require little entertainment (except, that is, when you settle down to write a book) and they exercise themselves, given the opportunity.

The prevalence of inherited defects is lower in cats than in dogs – although there is a depressing and enduring interest in a number of forms of cat with apparently endearing physical defects which are linked to much more serious conditions. For example:

- The Scottish Fold is a breed of cat that carries a genetic mutation which affects the development of cartilage. The most obvious (and apparently endearing) impact of the mutation is the way in which the ears fold forward towards the front of the head. As the mutation affects cartilage throughout the body, Scottish Folds develop abnormally short limbs, an abnormal gait and a peculiar stiff tail. A lack of cartilage increases the risk of early-onset osteoarthritis and severe lameness.[10]

- The Manx is a breed of cat with a genetic abnormality which is most obviously expressed by either the complete absence of a tail or merely the presence of a short stump. However, like the Scottish Fold, the effects of the mutation are often more profound. The mutation also affects the spine and spinal cord. This results in a form of spina bifida – a developmental abnormality of the spine that can result in problems with control of urination and defecation, and sometimes also with control of the back legs (causing partial paralysis).[11]

Again, outbreeding would rid the breeds of these problems within a few generations. It would also mean the loss of the endearing physical traits, but I'd argue that was a small price to pay for improving the health and welfare of our pets.

Of course, we suppress or seek to alter the behaviour of cats by preventing breeding, inhibiting hunting (with varying success), and discouraging them from marking their territories indoors – but again I would argue that these insults are less severe than those visited on farmed and laboratory animals.

Mutilations

Both dogs and cats are mutilated for our convenience. Tomcats are castrated and queens (female cats) are spayed to prevent them breeding, fighting, spraying urine, smelling and straying – all behaviours that are part of the normal behavioural repertoire of the cat. Dogs and bitches are also castrated and spayed respectively, again to stop them breeding and in an attempt to reduce aggressive behaviour in males.

The incidence of mammary carcinoma (breast cancer) in neutered bitches is lower than in entire bitches. Similarly, conditions of the uterus, some of which can be life-threatening, can be avoided by spaying bitches. But that is not the reason for the procedure. These mutilations, for that is what they are, are for the convenience of the owner, and any medical benefits are entirely secondary. I don't argue against spaying or castration. It is a useful and generally safe procedure. Anaesthesia is compulsory, unlike during most mutilations of farmed animals, and given the benefits to society as a whole, neutering both female and male dogs and cats has utility. I simply want the procedures to be recognised for what they are – mutilations – and hence acknowledged to be done for the convenience of the pet's owner, not for the good of the animal.

Many breeds of dogs are traditionally tail-docked, particularly the so-called working breeds. This is a contentious issue. Advocates argue that tail docking is necessary to prevent injury to the tail, conveniently ignoring that the dog has evolved a tail for a good reason. It is a fatuous argument, and one that has been debunked by a study which found that approximately 500 dogs would need to be docked in order to prevent a single tail injury.[12] One might also argue that the risk of injury to any bodily appendage can be reduced if you surgically remove it beforehand. That is, you could reduce the number of leg injuries if all dogs had at least one of their legs amputated.

To be clear, the removal of a dog's tail is an outdated practice. It inflicts significant pain on puppies and deprives dogs of an important form of expression. The BVA believes that tail docking should be prohibited. In most circumstances it is, but there are a number of exemptions and there has been some backsliding in Scotland.[13] In summary, like the breeding of dogs with flat faces, tail docking is a painful procedure which provides no benefit to the dog and serves no purpose other than inflating the vanity of the owner. It needs to stop.

Given the reduction in tail docking, it is profoundly dispiriting to see that another cosmetic mutilation is becoming fashionable. Ear cropping, a surgical procedure to remove part of a dog's ears so that they stand upright rather than flopping forward, is illegal in the UK. It is normally performed on Doberman Pinschers, Great Danes and Staffordshire Bull Terriers when they are between 6 and 12 weeks of age. Like tail docking, it serves no purpose other than to feed the vanity of the dog's owner. No vet will perform the procedure but a 'kit' containing the necessary instruments (but with no anaesthetic) can be bought online. Reports from vets in practice suggest that more dogs are being cropped. It is a painful, disfiguring and abhorrent procedure, and those people who perform the procedure need to be caught and prosecuted.

Obesity

Additional complications of pet ownership are the conditions that dogs, cats and other animals develop simply as a result of our meddling in their lives. For instance, obesity is a growing problem in dogs and cats. Overfeeding is as much a welfare issue as underfeeding or malnutrition. Obesity in dogs carries some severe welfare consequences, from associations with shortened lifespan to a higher incidence of arthritis, diabetes mellitus and certain types of cancer. It is easier to prevent obesity than to treat it. Indeed, the problem can be largely avoided by feeding the right diet in the right amounts. But the problem is getting worse,[14] and now we have veterinary clinics running wellness programmes for their clients' overweight dogs.

Exotic animals as pets

In addition to the commonly kept animals such as dogs, cats and rabbits, many other species are kept as 'pets'. Some of these can be described as domesticated, like the guinea pig and gerbil. Many, including primates and some species of reptile and amphibian, cannot. Domestication is much more than simply taming. It is the process of adapting a population of animals to a lifetime of captivity and dependence on human-mediated husbandry and the associated changes in behaviour. An animal can be described as domesticated when, after several generations of selective breeding, it has become adapted to life in close association with humans. Dogs and cats fit that description. Few, perhaps no, reptiles and primates do.

It is likely that meeting the needs of these exotic animals (or 'non-traditional companion animals' as the BVA describes them) will be difficult in many cases and impossible in others. I am inclined to the BVA's view that keeping these animals is such a specialised task that it isn't, in general, suitable for a household environment, particularly when the keeper is inexperienced. Those who do keep them should be certain that they are able to fully meet the animals' welfare needs.

Chronic conditions and geriatric pets

Pets are not people. Neither a veterinarian nor the owner of a pet can explain to a dog or a cat that the invasive surgery or that the long-term treatment about to be administered will prolong its life nor that that the pain and discomfort will be worth it. Increasingly, however, improvements in technology and treatment regimes mean that treatments for cancers, severe arthritis, thyroid dysfunction, diabetes and even paraplegia are available for pet dogs and cats, and they are increasingly used. This immediately raises an ethical question. Is it ethical to perform highly invasive surgery or prescribe long-term treatment for conditions which will merely extend life for a few weeks or months, knowing that the animal is likely to suffer prolonged discomfort and pain?

Companion animal vets have been wrestling with this dilemma for years and, as befits such a sensitive issue, there are all shades of opinion. My time in companion animal practice was admittedly short and many years ago. But then, mindful as I was of the way in which pets seem to suffer chronic discomfort stoically, I was inclined to advise euthanasia at an earlier stage than my colleagues. I'd ask myself, 'What sort of quality of life is this dog or cat going to have?' And if the answer was 'not much' and there was little prospect of any treatment changing this assessment any time soon, then I would advise accordingly. Of course, not everyone would take that advice, and that's their choice. There were times, perhaps, when I was less tactful than I have become now, and on occasion I would imply that an owner who wished to continue with treatment was simply being selfish. On one occasion, at the tender age of 23, I was told by an elderly couple, after I had advised euthanasia for their dog, that my experience of 'life' was insufficient to appreciate its sanctity. Given that the tumour on the dog's abdominal wall had grown so large it could no longer stand up, let alone walk, I felt my advice was sound. And since my employment as a *locum tenens* in the practice was coming to an end the next day, I felt able to explain, as bluntly as possible, why I believed they were wrong.

These decisions are not easy, and each case has to be assessed on its merits. It's hard if you have to say goodbye to a dog or a cat that has been your companion for a decade or more. But we have a responsibility that extends beyond our own immediate needs, and when it comes to the crunch, so to speak, then I believe that in many cases euthanasia sooner rather than later is preferable.

Keeping pets represents an ethical dilemma

Do the various ways in which we have an impact on the lives of our pets present an ethical issue? That is, does the keeping of pets require ethical scrutiny similar to what is required for farmed animals, wildlife and research animals? If we take into account the constraints we impose on the behaviour of our pets, the impact of breeding for undesirable characteristics simply to satisfy our vanity, the mutilations and the overfeeding (and perhaps enforced indolence) of dogs and cats, the argument seems pretty strong to me.

Pet owners may not eat or experiment on their pets, they may not hunt or trap them, but it is an unavoidable conclusion that pet ownership is a form of exploitation. This exploitation is primarily for our benefit, and it has a number of consequences. It is also an unavoidable conclusion that for a substantial but unknown proportion of pets the experience is far from optimal.

That is not to say that pet ownership is to be abhorred. Far from it. But recognising it as a form of exploitation with similarities to the way in which we treat and breed farmed animals is a step towards ensuring that efforts are directed to improving the way it is done.[15]

However, that is not a universally held view. While some would question whether the experience of pets is an ethical issue because of the care and attention lavished upon them, at the other end of the spectrum are those who argue that keeping pets is unacceptable on ethical grounds. In general, people who maintain that the breeding and keeping of pet animals is unacceptable believe that no animal, irrespective of its circumstances or its purpose, should be exploited, no matter how benign that exploitation appears to be. This is because, they believe, 'happy exploitation' is impossible.

All keeping of animals involves ownership, and all ownership denies freedoms to the animals involved. Others argue that keeping pets deprives animals of their rights. It is a rational position and one that it is difficult to argue against if you take the vegan ethos to a logical conclusion – although I note the efforts of some vegans to change their dogs' diet to a vegan one.

I sit somewhere in the middle in that I believe that it is acceptable for animals in a modern society to be exploited, so long as this is done within well-defined criteria. Those criteria need to be tightly drawn, preferably by societal consensus, and regularly reviewed.

In comparison with the worst excesses inflicted upon farmed and research animals, it might be argued that the impacts on pet animals are relatively benign. But it might also be argued that these impacts and interventions are so familiar that we barely notice them. This, particularly if you have strong opinions about the welfare of farmed animals, research animals and wildlife, ought to concern you.

It means that the keeping of pets, as well as providing you with the benefits of companionship and affection, confers upon you some responsibilities. These include the maintenance of good health and good husbandry, which among other things means providing pets with the opportunity to act 'naturally' wherever possible (while not causing a nuisance to other people and animals). It also means avoiding pets which are deformed through crass selective breeding and selecting only dogs that are dog-shaped, cats that are cat-shaped and rabbits that are rabbit-shaped. Finally, decisions about treatment, particularly for chronic conditions in older animals, must take into account a balance of benefits – thinking of the pet first and foremost rather than the wants of the owner.

I am an advocate of reducing the numbers of species that may be kept, of strict limits on the types and numbers of mutilations, of efforts to reduce the constraints on pet animals' behavioural repertoire, of vigorous efforts to eliminate undesirable inherited traits, and of more responsible pet ownership that addresses 'owner-induced' conditions such as obesity. In other words, we need to hold ourselves, as pet owners, to the same standards that we expect of any other exploiter of animals.

Whether or not you consider these to be 'rights' for pets is open for debate. I am not sure one way or the other. However, it is clear to me that any consideration about the rights of animals, in whatever circumstances we find them, has the potential to secure improvements in the way in which we treat them. You may not agree that a pug has the 'right' to breathe normally, nor believe it has the 'right' to walk 500 metres without becoming exhausted. You simply have to believe that it is unacceptable to breed an animal to meet a bizarre and selfish standard which results in these privations and indignities. Such a viewpoint might not go far enough to meet an animal rights agenda, but treating our pets with greater respect is an important element in the recognition of our responsibilities towards animals.

10

Animals Used in Research

My first brush with the debate about research animals happened when I was still in my teens. I went to a comprehensive school in Paisley. English was not my strongest subject and I was streamed in a class of similarly struggling but potentially salvageable 14-year-olds. We were taught by a dapper, eloquent man who tried valiantly but ultimately in vain to get us interested in the poetry of Robert Burns and all manner of worthy but obscure Scots literature.

One day, after reading aloud passages from *The House with the Green Shutters*, again, the teacher announced that next week we were to hold a debate with the motion 'This house believes that experiments on live animals are wrong'. Foolishly and without hesitation, I volunteered to propose the motion.

I went home fired with enthusiasm – only to realise something I should have known when I volunteered. I knew nothing of the subject. I did what every teenage boy does or should do when he is in a jam. I went to my mum. Ever resourceful, my mother contacted the headquarters of the Scottish Anti-Vivisection Society in Glasgow and arranged a meeting. On the next Saturday off we went, my mother and I, by train to the big city and to the top floor of a tenement block in West George Street. There we were met by an earnest, pallid man in a three-piece tweed suit that had seen better days. His office, which had also seen better days, was dusty and piled high with boxes and boxes full of papers, spilling out over the room.

We explained our interest, and there followed the most detailed and heartfelt argument against vivisection that I have ever heard. This was punctuated with a constant stream of leaflets and booklets that our host picked deftly from the overflowing boxes. I remember photographs showing a dog with two heads (one it had been born with, the other surgically attached), rabbits with the most appalling injuries, and serried rows of caged mice awaiting their terrible fate. It was horrible, compelling and quite upsetting. Or it was to me.

My mother, being of sterner stuff and having nursed throughout the Second World War, seemed largely unmoved.

The imparting of the information and the collection of the leaflets took some time, and before long we had missed three trains home. Eventually, when the gentleman finally paused for breath, we made our excuses and got away. I went home to prepare my arguments. But it was all too much – and two days later I was found in tears, clutching the photo of the dog with two heads, and was sent to bed early, overwrought.

I don't remember much about the debate other than the audience liking the leaflets, especially the one with the two-headed dog. My opponent, despite my earlier wobble, was a great deal less well prepared than me. She may have been mercilessly and clinically dispassionate about the need to use animals in science so 'they don't have to use you', but that was as far as it went. There were no photos of the contented subjects of vivisection and no leaflets. The teacher didn't ask for a show of hands, so I'll never know whose argument prevailed.

Was the debate and my preparation a formative experience? Probably not, but that and the other developing interests from my teens such as birdwatching did nudge me towards a career with animals. Possibly more important, it was my first experience of the power of lurid illustrations to help force home an argument. A photo of a dog with two heads is always going to be more powerful than the sober, rational views of a researcher, however many life-saving medicines he or she has developed. And as my knowledge grew and I discovered that decisions about animal protection, exploitation and welfare are rarely binary, the experience slipped into the depths of my memory – only to re-emerge when I became an independent member of an Animal Welfare and Ethical Review Body (AWERB) in one of the country's leading research centres.

The exploitation of animals for research

Many people would consider that animals used in research and science are the most exploited group of animals. Who would argue with that? It is all too easy to develop an exaggerated caricature of the entire enterprise. This one is mine:

> They take blameless creatures, bred in huge numbers expressly for research, house them in sterile, barren environments and then subject them to all manner of cruel and unnecessary procedures, after which they are disposed of unceremoniously. It's all simply to build the careers of heartless and uncaring scientists. The work is secret, poorly regulated and inadequately scrutinised.

Of course this is a parody, but it is illustrative of the charged nature of the debate. Supporters of animal research argue that it is essential. For example, the advocacy group Speaking of Research believes that 'animal research remains crucial to sustaining life and protecting the planet'.[1] On the other hand, Cruelty Free International, a group campaigning against the use of animals in science, believes 'there is no ethical justification for using animals in experiments'.[2] Both organisations go into considerable detail to support their respective cases. Mostly they make cogent and rational cases.

I am not going to fall on one side or the other of the argument, because it's not that simple. As with all animal exploitation, there is good and bad, justifiable and unjustifiable. And with animals in research, as with all animal exploitation, where the line gets drawn between acceptable and unacceptable is a matter of personal values and ethics. Perhaps, more importantly, those values change over time, and what might have been acceptable 25 years ago may well appear unacceptable now.

Before delving into the detail, here's a thought. Many people are passionately and vehemently opposed to research using animals. However, it is inescapable that research using animals has saved countless human lives and will continue to do so. It is a highly regulated environment, and its standards of husbandry and training, alongside its levels of scrutiny, reporting and accountability, put all other types of animal exploitation to shame. To be sure, we need to ensure effective alternatives are researched and adopted, and to apply the most rigorous standards for those animals we use. But that won't satisfy the abolitionists.

I understand why some people seek abolition, but, if one takes a utilitarian position, strictly limited and well-regulated use of animals in research can be justified because of the benefit to humanity. It's certainly easier to formulate a supporting argument than to create one to justify continuing to eat meat, no matter how much the rearing of the animals meets those animals' needs. That is because eating meat doesn't save lives, and because there are nutritious alternative diets.

Regulation and controls

Research using animals has a long history in Britain, as has the organised opposition. It also has a considerable record of achievement, albeit sullied by some unwholesome and unedifying events which risk damaging its reputation.

Animal research in the UK is highly regulated, as it is in the European Union. It is less highly regulated in other parts of the world, notably in the USA, where standards of housing, care and other protections are very different and, arguably, do not offer the same levels of protection that prevail in Europe.

Britain was one of the first countries in the world to introduce legislation to control animal experimentation. In 1875, opposition to vivisection led the government to set up a Royal Commission on Vivisection, which recommended that legislation be enacted to control it. The Cruelty to Animals Act 1876 was enacted as a result and instituted a licensing system for animal experimentation. The Animals (Scientific Procedures) Act 1986, often referred to as ASPA, was enacted 110 years after the previous Act and is widely recognised as providing one of the toughest regulatory environments in the world.[3] This legislation made substantial changes to the regulatory landscape:

- ASPA defined regulated procedures as animal experiments that could potentially cause 'pain, suffering, distress or lasting harm' to protected animals, which encompassed all living vertebrates other than humans.
- Primates, cats, dogs and horses gained additional protection over other vertebrates.
- ASPA regulates the modification of genes in protected animals if this causes the animal pain, suffering, distress or lasting harm.
- Other sections of ASPA regulate animal sources, housing conditions, identification methods and the humane killing of animals.

ASPA involves three levels of regulation, person, project and place:

- The **person** level is achieved by the granting of a 'personal licence' to a researcher wishing to carry out regulated procedures on a protected animal.
- The **project** level of regulation is governed by the granting of a 'project licence' to a suitably qualified senior researcher. The licence details the scope of the work to be carried out, the likely benefits and the costs involved in terms of the numbers and types of animals to be used, and the harm that might be caused to the animals. Applicants must explain why the research cannot be done through non-animal or *in vitro* methods.
- The **place** where regulated procedures are carried out is controlled by the granting of an 'establishment licence' to a senior authority figure at the establishment.

The 1876 Act was criticised at the time because it made no provision for public accountability of licensing decisions. Interestingly, despite the wide regulatory changes introduced, ASPA still has no provision for public accountability of licensing decisions, although the mandatory requirement for a non-technical summary with each application is a tentative step towards this.

Perhaps in recognition of this apparent lack of accountability, initial guidance on the implementation of ASPA recommended that each licensed premises institute ethical review processes to scrutinise project licence applications. When ASPA was amended in 2012 to implement an EU Directive, ethical review became mandatory.[4] Each breeding, supplying and user establishment was required to set up an AWERB. The AWERB acts to ensure that all use of animals in the establishment is carefully considered and justified; that proper account is taken of all possibilities for reduction, refinement and replacement (the 3Rs – see below); and that high standards of accommodation and care are achieved.[5]

As a concept and in practice the AWERB, being a body constituted locally to hold responsible people to account for animal exploitation, may be unique. Although central bodies such as the Animal Welfare Committee (AWC) exist to advise central government, there is no local governance capable of influencing the way in which farmed animals and other kept animals are exploited. And as for wildlife, there appears to be nothing at any level, although some conservation bodies hold themselves to account through ethics committees with external representation. Horse racing organisations have their own ethics committees but these are run at a national level. And perhaps if there was a degree of ethical scrutiny of dog breeding there would be fewer misshapen dogs around.

What research can be done?

The types of research that may be carried out are, theoretically, unlimited, and many of them might be considered unethical. ASPA, however, limits the types of research that can be licensed, and most of the approved types are applied. In plain words, applied research offers the prospect of a direct application to real-world problems: for instance, the avoidance, prevention, diagnosis or treatment of disease; the improvement of the welfare of animals; the development, manufacture, or testing of the quality, effectiveness and safety of drugs and foodstuffs. It can be argued that applied research of this type benefits society despite the use of animals.

ASPA also allows for basic research, that is research which is simply a quest for knowledge without an immediate practical application. Basic research accounts for a high proportion of the total number of animals used (see below).

Harm versus benefit

An important element of ASPA is the requirement for harm–benefit analysis. This is the process of determining whether the harms to the animals are justified by the potential benefits.[6] It is a complex process but in essence it is simply a method of determining the utility of the proposed work on animals. Harm–benefit analysis is undertaken by veterinary or medically qualified advisers with particular expertise in assessing research proposals.

Part of the process involves categorising the proposed procedures by their expected degree of severity. Procedures are classified as 'sub-threshold', 'mild', 'moderate', 'severe' or 'non-recovery', based on the degree of pain, suffering, distress or lasting harm that the animal is expected to experience.[7] The proposer is expected to develop procedures which keep severity as low as possible while not compromising the project aims. However, where the potential benefits are high a severe procedure is more likely to be accepted in the harm–benefit analysis.

The harm–benefit analysis is particularly difficult when considering proposals for basic research, as it is more difficult to determine the benefits when no immediate practical societal benefit can be expected.

The numbers count

The Home Office, the government department responsible for the regulation of animal experimentation, publishes annual statistics about the numbers of animals used in research.[8] It is a useful document in that it contains a detailed breakdown of the species used and the types of procedures and, perhaps most importantly, compares the figures with previous years. In contrast, there are no similar reports for the main groups of other exploited animals such as farmed animals, wildlife and animals in sport.

The Home Office's imperative is to reduce the numbers of animals used and to reduce the proportion of animals subject to 'severe' procedures. The statistics show that although there have been changes in the way the figures are calculated, over the last 25 years the trend is downwards.

In the mid-1970s numbers peaked at around 5.5 million animals. When ASPA came into force in 1986 there were around 3 million procedures annually. This fell gradually, year by year, to around 2.5 million by 2003. The numbers rose slowly back to around 3 million by 2012. The increase between 2003 and 2012 can be attributed primarily to the increasing use of genetically modified animals – which, although they may be simply breeding and multiplication stock, still count toward the total. As of 2020, the numbers of animals used had fallen by 15% in comparison with 2019 and represented the lowest figure since 2004. The proportion of severe procedures in 2020 was reported at 4%, and while this is not a reduction against 2019, there has been a downward trend over the last few years.

This all sounds like good news, but the headline figure of 2.88 million animals used during 2020 is hardly trivial. What exactly happens to the animals? And which species are involved?

Animals are used in four main areas of biomedical research and product testing:

- **Applied research.** Here animals are used in developing better medical products and surgical techniques. Much of this work is for human medicine; some is for veterinary use. Animal research plays an important part in the search for cures for chronic diseases such as multiple sclerosis, the various forms of dementia and certain cancers, the development of new medicines and vaccines and improving existing ones. Applied research accounts for around 10% of procedures.
- **Regulatory testing.** Chemicals in household products as well as those used in manufacturing, or as fertilisers and pesticides used in farming, must be tested to make sure that they are as safe as possible for animals and people. Regulatory procedures focus on safety and efficacy. The most common procedure in 2020 was toxicity and other safety testing, including studies for safety evaluation of products and devices for human medicine, dentistry and veterinary medicine. No testing using animals has been allowed on the ingredients of cosmetics or toiletries since 1998, and no testing of household products since 2015. Regulatory research accounts for 12% of procedures.
- **Breeding of GM laboratory animals.** Most genetically modified (GM) animals are used in genetic studies. GM mice and rats account for the vast majority of laboratory animal breeding. They may be used in basic research to discover the function of a particular gene, or in the study of diseases. GM mice are useful in the study of many diseases that are inherited or partially inherited, caused by faults in an individual's genetic code, such as potentially fatal illnesses like sickle cell disease or cystic fibrosis. The breeding of GM animals accounts for around 50% of animal procedures.
- **Basic research**, it is argued, is necessary to provide the knowledge that ultimately supports the applied research, and is analogous to the foundations of a house: if you haven't laid the foundations properly then the house will either fall down or be impossible to build in the first place. Basic research accounts for around 25% of animal procedures.

Given that basic research accounts for around a quarter of all animal use, the following question seems pertinent: How can a harm–benefit analysis on basic research be conducted if the likely benefits cannot be determined? I address this below.

The 3Rs: replace, reduce, refine

In 1959, William Russell and Rex Burch published a highly influential book, *The Principles of Humane Experimental Technique*, introducing the concept of the 3Rs to animal science.[9]

Russell and Burch defined the 3Rs as follows:

- **Replacement** means the substitution for conscious living higher animals of insentient material.
- **Reduction** means reduction in the numbers of animals used to obtain information of a given amount or precision.
- **Refinement** means any decrease in the incidence or severity of inhumane procedures applied to those animals which still have to be used.

The original definitions have been amended since the book's publication but the basic principles remain the same – use fewer animals, use animals with a lower capacity to suffer, use alternatives (for example, cell culture), and when it is unavoidable make the experience less painful and less unpleasant.

The principles of replacement, reduction and refinement now underpin all animal experimentation in Britain and throughout much of the world, to the extent that in Britain, unless the principles have been applied to experimental design from the outset, a project licence will not be granted. A publicly funded body, the National Centre of the 3Rs (www.nc3rs.org.uk), awards funding for research, innovation and early career training to a wide range of institutions to accelerate the development and uptake of 3Rs approaches.

Which animals should be used in scientific research?

This is a debate that rumbles on and on. Assuming that you are not implacably opposed to all animals in research, where do you draw the line about which species can be used? In the past, this debate was bound up with the now discredited but still widely used 'continuum of development'. This Platonic concept is explicit in the idea of the 'great chain of being', or a 'progression of perfection': all beings on Earth, animate and inanimate, could be organised according to an increasing scale of perfection from, say, bacteria at the bottom up through lobsters and cattle, all the way to human beings at the top. As the theory of evolution through natural selection became widely accepted it became clear that this notion was nonsense. There is no continuum; there is no steady progression from the 'primitive' to the 'perfect', and no parallel progression in complexity and the capacity to feel pain. However, the belief in linearity persists, perpetuated by depiction on T-shirts (showing a quadrupedal primate on the left and a bipedal human on the right with progressively more upright primates in between) and by complacent journalists.

How does this impinge on animal research and the choice of species? At its crudest, it means that the care and consideration given to species assumed to be 'well down the hierarchy' such as fish and invertebrates is not particularly lavish. It was not until an amendment in 1993 to the current Act (ASPA) that the common octopus was brought within its scope,[10] and not until 2012 that other cephalopods were included.[11] Decapods, despite mounting evidence of sentience, are still not included.

A principle embedded in the implementation of ASPA is that any project licensed under the Act must 'involve animals with the lowest capacity to experience pain, suffering, distress or lasting harm'.[12] Taking this well-established principle into account when designing the project ensures that applicants can justify why a particular species was chosen to answer the scientific question. On the face of it, that seems quite simple: from a list of species where there is evidence that valid results will be obtained, select the one least likely to suffer while the study is under way. If, for instance, a toxicological test can be conducted using *Daphnia* (a small aquatic crustacean) instead of a fish, there can be little doubt that this would meet the requirement. It becomes more difficult when the choice of alternative species is limited to mammals.

Remember that in Chapter 2 I concluded that, because of the difficulty in determining whether an animal has the capacity to suffer and in the absence of contrary information, it was prudent to assume all vertebrate animals have that capacity. Therefore, the decision as to whether, say, a rat's capacity to experience pain is less than a cat's is not going to be informed by the science currently available. More likely it will be influenced by emotional preconceptions regarding biological characteristics and how society values one species over another. While there are undoubtedly other factors at play such as cost and lack of suitability, the very low numbers of cats (and dogs, horses and primates) used in animal experimentation may simply be a reflection of society's emotional response to these species.

A further pressure bearing down on these numbers is the additional protection afforded under Section 5C of ASPA, which makes these four animals 'specially protected'. Before they can be used, licence holders must demonstrate that no other species are suitable, and they must adhere to additional licence conditions. In 2020, fewer than 1% of the animals used were cats, dogs, horses or primates.[13] There are further constraints which apply to all animals within the scope of ASPA, but in practice primarily to cats and dogs: a presumption against the use of animals not expressly bred for the purpose and against the use of stray and feral animals.[14]

The use of primates in animal research is, if anything, even more contentious than that of dogs and cats. This is partly because the use of primates often has an added dimension. Much of the work is neuroscience – investigating the architecture, function and interconnectedness of the central nervous system – requiring lengthy periods of training, invasive surgery and sometimes a period of several years over which experiments are conducted.

Primates are used specifically because they have very similar brains to ourselves. In fact, the brains of the great apes (gorillas, chimpanzees, bonobos and orang-utans) are the most similar to that of humans, and yet they aren't used. Why? Although when first enacted ASPA did not explicitly prohibit the use of any of the great apes, none has been the subject of a project licence. The UK government made it plain in parliamentary answers that it would not do so.[15] Calls for a formal prohibition were finally answered when the revised EU law on animal experimentation introduced a ban.[16] Other than New Zealand, no other country has so far followed suit.

The rationale for the prohibition in EU law included the following:

> The use of great apes, as the closest species to human beings with the most advanced social and behavioural skills, should be permitted only for the purposes of research aimed at the preservation of those species and where action in relation to a life-threatening, debilitating condition endangering human beings is warranted, and no other species or alternative method would suffice in order to achieve the aims of the procedure.

As with the restrictions on the use of dogs and cats, the decision to introduce a ban on the use of great apes is primarily an ethical one, and while there is a potential exemption, the bar appears to be set very high. It has not been universally welcomed. While most scientists appeared comfortable, some, like the late Professor Colin Blakemore, the then chief executive of the Medical Research Council (MRC), was concerned that a blanket ban might hinder research into important viral diseases and that British researchers might lag behind in other research where great apes were an essential element, such as understanding the roots of human language, social behaviour and self-identity.[17]

Those opposed to the use of animals in research welcomed the ban on the use of great apes. But they continue to campaign vigorously for a ban on the use of all primates. Which species of primate are used? There are two main species involved: the cynomulgus monkey (or crab-eating macaque) and the rhesus macaque. Smaller numbers of marmosets and tamarins are also used. Together, in the UK, primates are represented in around 0.1% of all procedures.

In comparison to the numbers of rats and mice, the numbers are tiny. And yet the controversy continues, with the government's statutory advisory group, the Animals in Science Committee (ASC),[18] having to examine in detail the arguments for continuing using primates in research, and the scientific community having to defend its position. An unequivocal answer has yet to be reached, since the question is essentially one of ethics: do the benefits to science and ultimately society outweigh the undoubted intrusive, repeated and protracted nature of many of the procedures? Inevitably, there is no simple answer.

After leaving public service, for several years I sat on an AWERB as an unpaid lay member. The institution involved has a history of scientific achievement particularly in human medicine. The responsibilities of a licensed premises and of the people authorised to work there are considerable. These were taken seriously. Much of the work involved using mice and rats to conduct research into diseases and conditions that beset modern humans – chronic kidney disease, carcinomas, diabetes, coronary heart disease, the consequences of organ transplant surgery on the recipient's body, etc. While we were concerned to ensure that the 3Rs were rigorously applied and that there was continuous improvement, there were few fundamental objections to this type of research. It became more difficult, for me at least, when discussing proposals to use primates. I was prepared to accept the use of primates to help develop vaccines for devastating infectious diseases such as dengue fever provided there was evidence of compliance with the 3Rs, particularly in the areas of refinement and reduction. Since an effective dengue virus vaccine would save thousands of lives, there was a positive harm–benefit analysis. I was less comfortable with the use of primates for neuroscience.

Much neuroscience involving primates is basic science. The knowledge gained might have a practical application at some stage in the future, but the primary motive of the principal investigator, the institution that hosts the research and the body that funds primate neuroscience is the accumulation of knowledge. Neuroscience research has a timeline over decades, dedicated to uncovering the mysteries of the central nervous system. The techniques have been refined over the years, to the extent that the functions and connections of individual nerve cells (neurons) can be studied. But these techniques are invasive: access to individual neurons requires surgery and permanent access to the brain. The primates have to be trained to perform tasks and are often involved for months and sometimes years. While it is clear that standards of care are high, the fact remains that these animals are exploited for a prolonged period. So much so that a debate rages about whether they experience 'cumulative harm' or 'cumulative severity'. The ASC has been grappling with this for years and has yet to reach a definitive view about the harms and benefits of neuroscience using primates. Perhaps it is impossible. For further reading I recommend a report on the cumulative severity of neuroscience research in non-human primates[19] and the ASC's response.[20]

There is something missing from the debate. There is no doubt that despite excellent care and efforts to create interesting and social environments, the experience of these animals is poor. A utilitarian approach would accept the harm if there were tangible benefits. There are clearly some benefits to the investigator and the hosting institution. But, as the work is likely to be funded out of taxation, one might ask where's the benefit to us, the taxpayer, the citizenry? Given that any practical application could be decades away, that question is impossible to answer. However, history might come to our aid.

Parkinson's disease (PD) is a debilitating and distressing neurodegenerative disease. The quality of life for an increasing number of PD patients has been improved by the development of techniques which relied on basic research into the neural control of voluntary movement in monkeys. Research over many years had pinpointed the parts of the monkey brain responsible for control of voluntary movements. This knowledge aided the development of deep brain stimulation (DBS) to effectively alleviate severe dyskinesia (uncontrollable movement) in advanced PD patients, and the treatment has proven extremely effective.[21] Around 10% of PD patients suffer most from the opposite problem, akinesia (loss of ability to create muscular movements), and here too, work on primates, again building on the basic knowledge of the functional architecture of the monkey brain, has aided the development of effective treatments based on DBS.[22] It is very unlikely that such treatments would have been developed without prior knowledge of the monkey brain. While this does not solve the ethical dilemma, it does demonstrate that we dismiss at our peril basic research where the immediate utility is hard to demonstrate.

Even if one accepts that basic research is essential, there is always a point at which we, the citizens, should have the opportunity to say 'enough'. I like to think that the decision to stop research on great apes reflects the views of society. There is evidence that supports this view, and while there are some advocates for their continued use they remain very much in the minority, with many scientists citing moral arguments against the use of these species.[23] However, while I respect many of the scientists involved in neuroscience, I would be staggered if any of them had voluntarily offered up a ban. There is a lesson here: change comes from public pressure, scrutiny and regulation rather than waiting for voluntary measures.

Of all the ethical arguments associated with animals in research, the use of primates is the one that evokes the greatest passion. But as Chapter 2 explains, we need to be careful in classifying animals by their 'worth' or by how society values them. As the evidence builds of the capacity of vertebrate animals (and some invertebrates) to suffer, the argument that we should manage our relationships with animals on the basis of their needs is strengthened. It is likely that needs of primates are greater and more demanding than those of rats, mice and fish, but this does not mean that those species are less deserving of proper care and protection than primates. Thankfully the law recognises this, as does the scientific community.

Alternatives to using animals in research

Proper application of the 3Rs means developing alternative systems in which animals are not necessary. There are a number of alternatives in current use and more are under development. Examples include the following:

- **Cell cultures.** Cell culture is the process by which cells are grown under controlled conditions, generally outside their natural environment (*in vitro*). After the cells of interest have been isolated from living tissue, they can subsequently be maintained under carefully controlled conditions. Almost every type of human and animal cell can be grown in the laboratory. Cell cultures have been central to key developments in areas such as cancers, sepsis, kidney disease and AIDS, and are routinely used in chemical safety testing, vaccine production and drug development.
- **Human tissues.** Both healthy and diseased tissues donated from human volunteers can provide a more relevant way of studying human biology and disease than animal testing. Human tissue can be donated from surgery (e.g. biopsies, cosmetic surgery and transplants). For example, skin and eye models have been developed based on donated material, and have been used to replace animal models.
- **Computer models** of the heart, lungs, kidneys, skin, digestive and musculoskeletal systems can be used to conduct virtual experiments based on existing information and mathematical data.
- **Volunteer studies.** Advances in technology have led to the development of sophisticated scanning machines and recording techniques that can be used to study human volunteers safely.
- **Organ-on-a-chip.** Human cells have been used to create innovative little devices called 'organs-on-a-chip' or, more accurately, 'organoids'. Organoids grow *in vitro* on scaffolds (biological or synthetic matrices) or in a culture medium and mimic the structure and function of different organs, and these can be used instead of animals to study biological and disease processes, as well as drug metabolism. Devices have already been produced that accurately mimic the lung, heart, kidney and gut, and these have replaced animal models for some types of research.[24]

But this is not the end of research using live animals. An organoid is *not* an organ in a dish. It is a simplified replica with some of the features of the organ it models. Artificial systems that mimic real organs could provide a better alternative for understanding mechanisms underlying physiological responses than current cell-based models or animal tests.

Regulators are keen that these alternative systems are widely adopted, as this will reduce the numbers of animals used. However, as you might expect, there are limitations – for instance, the study of disease dynamics in organisms where interconnectedness of the various vital organs renders even the most complex organ-on-a-chip unable to reproduce how the whole animal will respond. Live animal research will be with us for a few decades yet.

Replacing 'higher animals' with 'lower animals'

Stimulated by the principles of the 3Rs and in response to ASPA – which states that procedures must use 'animals with the lowest capacity to experience pain, suffering, distress or lasting harm' – there is increasing interest in using animals which are generally known as 'lower vertebrates' to replace 'higher' animals. This is because they are believed to have to have a lower capacity to suffer.

For example, the zebrafish has become a popular model organism in biomedical research. Multiple advantages of using this species include high physiological and genetic similarities to mammals, external fertilisation, rapid development, transparency of embryos and larvae, and the relative ease with which genetic and other experimental manipulations can be carried out.

The zebrafish is emerging as a new important species for studying mechanisms of brain function and dysfunction. The zebrafish represents an ideal organism for brain imaging studies because its small size and optical transparency enable imaging access to the entire brain at cellular and even subcellular resolution.[25] Counterintuitively, zebrafish possess high physiological and genetic similarities to humans and have become increasingly useful in studying a wide spectrum of human brain disorders, from brain cancer to epilepsy. A review published in 2014 concluded that 'although human behaviour will never be similar to fish responses (and vice versa), the evolutionarily conserved nature of complex central nervous system traits suggests that many human and zebrafish phenotypes share common genetic and physiological factors, representing an exciting emerging field for further translational studies in neuroscience.'[26] It is worth noting, however, that our understanding of how to meet the needs of zebrafish is not good compared to rodents and primates, and that zebrafish models involve the use of many more animals than rodent or primate models.

It is probably too early to say, but these developments may offer the prospect of reducing or even eliminating primates and other mammals from some if not all areas of neuroscience.

Concluding remarks

Despite starting over 150 years ago, the debate over animal research is not over. The abolitionists are not mollified by the most complex and onerous regulations, rigorous scrutiny and inspection, nor by ongoing efforts to reduce, replace and refine. They want it to stop.

On the face of it, there appears to be a rational argument for abolition, particularly when the utility of the exploitation can be effectively challenged and at a time when alternatives to the use of animals appear to be increasingly credible. It is clear, however, that this is an overly simplistic position. The need to conduct whole-animal work for basic research, and to study the systemic effects of complex disease, remains important. There is a strong utilitarian argument in support of work using animals to investigate, for instance, diseases of the central nervous system such as various forms of dementia and multiple sclerosis, where the alternatives do not yet mimic the complex functional architecture of the brain. Similarly, where the impact of chronic disease on a single organ such as the heart or the kidney has far-reaching impacts on the whole body, the likelihood of developing alternative systems that can reproduce these complex impacts in other parts of the body seems remote.

Efforts to develop technology that enables us to apply the 3Rs effectively need to continue, and scientists must be held rigorously to account at all stages of the application and review of project licences by both the Home Office inspectorate and the local AWERB.

What about basic research? Surely we now know enough about how the human body works? Why do we need to find out more? I wish it was as simple as that.

No matter how tempting it is, prohibiting or severely limiting basic research involving animals because there is no immediate practical application misunderstands science and the scientific method. Knowledge gained from basic research is the bedrock of effective applied research. However, because harm–benefit analysis of basic research is so difficult, proposals for basic research need to be rigorously scrutinised. Merely because technology enables a new line of research doesn't justify its deployment, especially if it is particularly invasive, protracted or likely to create cumulative harm, physical or otherwise.

Although AWERBs have been around for 10 years, the implied promise that these bodies would provide a window for the citizen into the use of animals in research does not seem to have been fulfilled. Most of the appointed lay members are not sufficiently detached from the work and the institutions to be effective. More could and should be done to 'lift the lid'. In Chapter 12, I explore how citizens could be better involved and more influential in the use of animals for research, and indeed in many other aspects of animal use.

11

A Personal Ethical Framework

The preceding chapters provide a basic walk-through of this complex and challenging subject. This book is not intended to be a comprehensive examination of all the different ways in which animals are exploited. There are plenty of sources of information covering the many facets of animal exploitation; the references and suggested further reading at the end are a good starting point.

This information, with a bit of thought and reflection, could be used by anyone to develop a personal ethical framework. One option would be to consider your position, species by species and farming system by system. You might conclude that all or none is acceptable. Or you might choose to accept the exploitation of some species in some circumstances but not in others. My only plea is that you seek out and use the most comprehensive and reliable information to inform your choices. It is all too easy in modern times to be swayed by siren voices and exaggerated claims.

Getting started

How do you make those informed decisions? Where do you start? The table on the following pages shows my own ethical framework, including justifications for the positions I've taken. It's a work in progress rather than a fixed blueprint, but it might be useful as a template. I certainly don't claim perfection: yes, there is science behind it, but it is also driven by personal choices and my own (unwritten) ethical code. I've limited it to food but it could conceivably be used for other animal products such as leather or down.

In the table, I've simply listed various animal products, and alongside each I've added, over two columns, the rationale I have used and the decision reached.

Which animal products should I eat? An example of a personal ethical framework

	Considerations	Decision
Poultry meat	I am far from convinced that any large-scale poultry meat production has good welfare. Some of the lower-input/lower-output system which use slower-growing strains offer better conditions. The welfare of the birds in intensive broiler and turkey production systems is poor. Substantial numbers of poultry are slaughtered by dubious means.	I rarely eat poultry meat at home unless it is from a local producer that I trust or it is guinea fowl (a bird that does not lend itself to fast-growing or intensive production). The same is true for restaurants, because of a lack of information about provenance.
'Game' birds	I don't doubt that some walked-up and rough shoots adhere to satisfactory ethical standards, but the birds cannot be distinguished from birds from industrial shoots. Impacts on biodiversity and the environment unproven but likely to be significant on intensively managed shoots. There is a lack of data about the welfare of shot birds, particularly about those wounded rather than killed. Welfare in rearing systems is often poor. A significant proportion of the people involved, especially in the uplands, have a poor record of compliance with wildlife law. Birds and mammals considered to be 'pests' are killed inhumanely.	I don't eat meat from 'game' birds.
Beef	Beef rearing systems vary enormously. Some have good welfare, others don't. Some aid biodiversity, many don't. The lack of verifiable information about provenance available to consumers makes buying difficult. Some systems benefit wildlife, especially wading birds – but, faced with 500 g of beef mince, how do you know? Reducing red meat consumption is beneficial for health reasons. Carbon footprint, irrespective of CO_2 sequestration in pasture and soil, exceeds that of all other meats. There are too many routine mutilations.	Despite enjoying beef, I am committed to eating less, and from sources that reflect my values.
Dairy	I worry about dairy. High culling rates, calves removed from dam at an early age and damage to the environment are major concerns. I am investigating local producers who claim exemplary standards. There is too little information about the provenance of milk. While I am convinced that the welfare of dairy cattle and their calves can be good, too few farms can demonstrate this.	I still eat and drink non-trivial amounts of dairy produce. This is under review.

Pork	Even when sows and their piglets are reared outside, 'finishers' are often confined in barren environments in the last few weeks before slaughter. Many sows are confined for weeks at a time. There are too many routine mutilations. I am convinced that certain systems of pig rearing, primarily outdoor systems, provide good welfare.	I buy pork from my local butcher, who is supplied by a free-range producer.
Lamb	It is difficult and sometimes impossible to get information about provenance and rearing. See my comments on beef. In many places, upland sheep rearing is preventing the restoration of native woodlands.	Despite enjoying lamb, I find myself eating less and less.
Eggs	Free range means everything from six hens scratching about in your back garden to units with 2,500 birds per hectare. Free-range units have their drawbacks – health, hygiene, etc. are poorer than in caged systems. Free-range birds are better able to express the normal range of behaviour. Subjectively the benefits of free-range systems, despite some substantial drawbacks, appear to provide a better experience than modern caged systems.	I only buy free-range eggs.
Venison	I am far from convinced that the accuracy of shooting, and hence the time to irreversible unconsciousness, is as good as it should be. I am convinced that the welfare of properly stalked and accurately shot deer can be protected.	I rarely get the opportunity, but I do eat venison when I can find it.
Fish	Evidence that fish are sentient is disputed but it is probably better to give them the benefit of the doubt. If one accepts that fish are capable of suffering, there is no method of catching them that does not involve some form of suffering. Arguably line-caught fish suffer less than those caught by commercial fishing in nets, but the data are scant.	I remain undecided.
Crustaceans	It is almost impossible to determine how these animals have been killed. Many crustaceans are not humanely killed. Instead they are boiled, frozen or dismembered while still alive. Some supermarkets and restaurants have issued statements in which they claim to use humane electric stunning equipment.	I'm still trying to get useful information. There is even the prospect of a change in the law which would outlaw inhumane killing of crustaceans. In the meantime, I am not eating lobster or crabs.

183

Of course, you don't need to be so formal and write everything down, but I find the process helps. In the absence of data that allow an objective comparison between the many different types of exploitation, it is impossible to provide rankings. Despite that, I believe it is possible to identify the best and the worst. Even if decision making is not perfect, altering behaviour to favour the more welfare-friendly foods (and other products) can certainly help to make a difference.

Can you trust the label?

Suppose you've completed your personal ethical framework. How do you put it into practice? Assuming that visiting each and every farm that vies for your custom is impractical, you will need to take other people's word for it. Unless you are buying in bulk or purchasing unpackaged food, you will be relying on information from the manufacturer or the retailer. In effect, the words on the label.

Labelling for packaged food is becoming more and more complex. There are statutory requirements – for example, the name of the food, a list of the ingredients including certain additives, with those ingredients known to be potentially allergenic given prominence. The weight and 'best before' or 'use by' dates must also be included. Depending on the nature of the food, places of origin and processing might need to be added. For some foods, information about storage and cooking instructions are mandatory. There is also a requirement for a limited amount of nutritional information. In addition, the manufacturer may choose to state whether the food is suitable for vegetarians, vegans or the lactose-intolerant, whether it is gluten-free or free of other allergens, whether it includes organically grown ingredients, and there may also be information on how to recycle the packaging.

All this makes for a busy label. It can be confusing for the curious consumer. Where's the information about animal welfare? You might be surprised, but there is almost nothing required on a statutory basis.

There is mandatory labelling of eggs which covers animal welfare. Eggs for sale in the EU and, at least for the time being, in Britain must be described according to the system in which the hens are kept – on free range, in cages or in barns. A degree of official scrutiny helps to ensure compliance to the extent that some farmers have been successfully prosecuted for passing off eggs from caged laying hens as free range.

Most other 'welfare labelling' does not have legislative backing. Systems like the National Farmers Union's Red Tractor and the RSPCA Assured scheme require farms to meet additional standards, with the produce labelled appropriately. In contrast, the additional animal welfare require-ments of the organic farm standards, for instance the Soil Association, have the force of law in that each farm is independently certified and periodically inspected. More niche producers sometimes provide detailed labelling, but unlike the formal standards, this is supported by little or no independent verification.

It is important that you don't read too much into these labels. Labels and certifications are not all-encompassing. For example, while the welfare on organic farms is generally of a higher standard than the statutory minimum, it takes little or no account of wildlife. Many dairy farms with organic certification are participants in the official badger killing programme. Similarly, seals may be killed around fish farms which are part of the RSPCA Assured approval system.

Would there be value in compulsory welfare labelling? A 2018 government consultation found that 72% of respondents were in favour of the government setting further standards to ensure greater consistency and understanding of welfare information at the point of purchase.[1] In response to this, in late 2021, the government put out a call for evidence about labelling food with more information about animal welfare.[2] Respondents were asked for their views about labelling which provides information about rearing systems, the methods of slaughter, whether it should apply to all food or only home-produced food, whether the labelling should be compulsory or voluntary, etc. It is a brave move. I don't say this because I oppose better information about food – far from it. But it is an immensely complex area. Respondents were faced with 55 separate questions, and were asked to provide detailed evidence to support each response – an indication of the many factors that will need to be taken into account in deciding how to label food. The consultation is now closed and deliberations continue. Any system will need to enjoy the confidence of consumers. This will be achieved only if the system is robust, compliance is assured and there are effective sanctions for miscreants. Perhaps this is coming: the recently published Government Food Strategy contains a commitment to 'consult on proposals to improve and expand current mandatory labelling requirements'. The strategy goes on to state that 'proposals will cover domestic and imported products [...] and will help consumers identify when products meet or exceed our high UK animal welfare standards'. This represents a considerable change and, provided they are backed with robust compliance measures, these proposals have the potential to assist consumers in making better-informed choices about the food they buy.[3]

The proportion of meals we eat outside the home or buy in to eat at home increases annually. It is even more difficult to determine the provenance of food in a restaurant or from a takeaway than it is for food you have bought and then cooked yourself. Have you ever asked about the way the chicken on the menu was reared? And been content with the answer? No, nor have I. However, things are changing a little and some fast food and restaurant chains are providing information about the welfare of the animals that are used to supply their outlets.

What questions might you ask your butcher or the waiter in the restaurant? Here are a few:

- How was this chicken reared? Is it free range? Is it one of the slower-growing breeds? At what age was it slaughtered?
- Is the pork free range? You say the pig was bred outdoors, but was it housed at any stage?
- How did you kill this lobster?

You will require a bit of chutzpah to carry it off. You don't want to make a fool of yourself. Nor do you want to humiliate the butcher or the waiter. You will need a bit of foreknowledge to frame the questions and to interpret the responses, particularly if the butcher or waiter is well informed. And to avoid embarrassment you probably ought to warn the members of your party that you intend to grill the waiter.

Animal products other than food

What about other animal products such as leather or duck down? Try asking the salesperson about the provenance of those new leather shoes when you are next trying them on. Let me know how far you get. I try on occasion, but in most cases the sales staff know so little that they will either be baffled about why you are asking or try to obfuscate. Asking becomes more of a sport than an effective exercise in gaining useful information. But it shouldn't stop you from trying.

Years ago, after inspecting a dozen or so American poultry slaughterhouses on behalf of the EU and finding myself with a few hours in Washington DC before the flight home, I wound up in a posh shopping mall. I was idly browsing in a shop selling expensive and frivolous shaving gear and inspecting a particularly soft shaving brush when one of the shiny salespeople came up and asked if I needed help. As I was feeling ebullient after completing the 'mission' and not a little mischievous, I asked about the provenance and welfare of the badgers that had supplied the bristles for the brushes. 'Oh, no,' she gushed, 'you needn't worry about that. They're not *American* badgers.' I had no comeback to that but I do now have a beard. I found out recently that most badger hair used in shaving brushes is collected from badgers farmed and killed – and according to People for the Ethical Treatment of Animals (PETA), the prevailing conditions in the mainly Chinese farms are poor and hence the welfare of the badgers is poor.[4]

Putting it into practice

It is inevitable that, if you change your diet and your buying habits, you will become the subject of interest to your friends – and perhaps you'll feel the need to explain. Here are a few tips:

- **Don't evangelise.** Especially at social events. By all means explain, but don't think you have to convert the world. No one wants a sermon at a party. Alternatively, try blogging.
- **Keep it under review.** Be prepared to change your mind when new facts or better information emerge. I say better information because you might want to reverse a decision not to eat, say, pork when a local producer starts to sell pork from a system that you are comfortable with. Or stop wearing leather shoes when someone produces all-synthetic shoes which don't look like something your five-year-old child would wear.
- **Consider things holistically.** Decisions like this are rarely black and white, and a number of factors need to be taken into account. For example, avoiding beef might help reduce greenhouse gas emissions and might be better for your health, but this has to be set against the value of some grazing systems for biodiversity. It might be better to eat grass-fed beef that has been reared on wood pasture, on seasonal wetlands or some other type of conservation grazing system instead of beef from a feedlot. It will be more expensive, perhaps quite a bit more, so eat it less frequently and in smaller quantities.
- **Don't flagellate yourself.** By all means stick to your beliefs, but don't be a martyr. I choose not to eat pheasant, grouse or partridge but I am not going to crumple or stomp out if I find the pâté I ate at that party contained pheasant. Be aware, try your best, but above all be practical.

Despite that, I encourage you to err on the side of caution. If you are buying ready meals or eating in a restaurant and there is no information about the provenance of the meat, I'd avoid it. Or, better still, ask.

Decisions about animal exploitation should not be taken in isolation, and our choices should be about more than just animal welfare. There are other factors I urge you to take into account. I am expert in none, but for what it's worth, here's my take:

- **Health.** Evidence is mounting that eating large amounts of meat and dairy products is not good for you. This applies particularly to fresh meat containing large amounts of saturated fats and to meat products such as bacon where nitrites and nitrates are used as a preservative. Red meat is particularly singled out as harmful.

- **Global heating.** Livestock farming is a net contributor to greenhouse gas emissions and we are being encouraged to eat less meat as part of managing climate change. However, some meats have a smaller carbon footprint than others. Beef is particularly high, while poultry meat is relatively low. In essence, the larger the animal and the longer it takes to mature, the more greenhouse gas per kilogram of animal product. And monogastric farmed animals (poultry and pigs), because of their simple digestive system, produce less emissions than ruminants (cattle and sheep).

  It is argued that some of the carbon footprint from lamb and beef produced from permanent grassland can be offset against the carbon sequestered in the soil or in other parts of the farm such as woodland. However, in comparison to the woodland which pastureland has replaced, this is trivial. And while this might lead you to suppose that pork and poultry are a better bet, bear in mind that the area of land devoted to producing their feed is continually growing, and that much of it is derived from former rainforest and other woodland.

- **Environmental impacts.** All farmed animals produce urine and faeces. the more animals in a given area, the greater the potential problem. It has to go somewhere, and it follows that there are particular risks associated with intensively managed units where the entire operation is contained in a series of buildings or where large numbers of animals are housed permanently or for months at a time. Some large dairy herds and large broiler units appear to have particular problems with pollution.

- **Biodiversity** is not something you would normally associate with a discourse about animal exploitation. But particular types of grazing are good for bird life. For example, without grazing, four species of breeding waders – snipe, curlew, redshank and lapwing, the populations of which are in free fall – would almost certainly disappear from Britain. Wild roe and red deer do some grazing, but there is nothing like a cow for removing large amounts of grass and other vegetation. And, because they allow for the free expression of most normal behaviour, extensive grazing systems are among the best for animal welfare. I should make it clear that I do not hold the same attitude to sheep, at least those grazing the uplands. There are many reasons why much of upland Britain is devoid of native woodland. Most of them are sheep.

It soon becomes clear that there are frequent conflicts between all these different factors. Those systems that are among the best for welfare (extensive cattle and sheep production), which in some cases support high biodiversity, have the worst carbon footprint, while pig and poultry production, at least those with any commercial prospect, based on intensive production of fast-growing animals, have the worst welfare but the lowest carbon footprint. There are no easy answers.

I've thought about becoming vegetarian or vegan and haven't yet reached a decision. Perhaps part of the reason why I haven't become vegan is that it would feel like a repudiation of my life to date and my whole career. I happen to enjoy many types of meat, but that shouldn't be the sole reason to carry on if the evidence against meat consumption continues to pile up. Conversely, if there is sufficient, robust information about good livestock production systems, with good evidence that animal welfare is protected, then why not?

Other than complete abstinence, the only reasonable conclusion is to eat less and better meat from sources you can trust. Obtaining information that engenders that trust is difficult, but for meat you cook at home it is not impossible. Ask – and if you are aren't happy with the answer, err on the side of caution. The same sentiments apply when you are eating out. Whatever your decisions, do your best to ensure that they are based on reliable information, and are consistent with your personal ethical framework.

12

Making Sense of It All

In this final chapter, I want to make sense of our complex and complicated exploitative relationship with animals. The relationship is necessarily complex, because these are complex organisms with complex behaviours. It is a complicated relationship not because it has to be but because we make it so, via our inconsistent norms, values and legislation. It could be simpler, a lot simpler. It could also be a lot better – and by that, I mean we could treat animals better, giving greater priority to their needs and wants.

Our relationship with the animals we exploit needs to be reset so that humane treatment and a recognition of sentience is always at the centre of our thinking and our actions. This chapter explains what we can do to help change that relationship – by acting as intelligent and committed consumers and as active citizens. The engines of change are science and changing societal attitudes – which, in turn, drive revisions to policy and legislation. Through this we can make a difference to the lives and experiences of millions of animals.

During my career as a veterinarian, I had the benefit of being able to 'lift the lid' on all manner of animal exploitation both in the UK and overseas. I haven't always liked what I've seen but neither was it universally awful. Animal exploitation covers a multiplicity of human-mediated activity, from the relatively benign to the downright objectionable and everything in between.

The risks to animal welfare from some types of animal exploitation are well evidenced and well managed. For much of the rest, it is neither of these. Some sectors are well regulated and some are not; simply, there are too many activities that are exempt from any form of effective control. Some arouse public interest, but many do not. Much of what goes on is hidden from public view, protected by private and state security – a case of out of sight, out of mind.

190

In some cases, information is freely available, and sometimes highly visible, but where animal exploitation takes place behind closed doors, public accountability is often scant. Shining a light on these activities will help in resetting our relationship with animals.

Arbitrary groupings perpetuate inconsistent treatment of animals

In the preceding chapters, I have – for convenience rather than because I believe in its validity – split exploited animals into several groups: farmed animals, wildlife, animals used in sport, pets and research animals. These groupings are pragmatic, follow current practice and reflect tradition. They are in no small part maintained by a complicated legal framework which has grown up piecemeal around them.

These groupings evolved decades ago when attitudes were different. And because they allow animal keepers, policy makers, legislators and enforcement bodies to apply simple but arbitrary norms and standards to one group at a time they became permanently codified and a matter of routine. It ought not to be that way. Sure, we may need to classify animals in a particular way so that we can make the right decisions on care and protection. But this ought to be based not on how we exploit them, but rather on an evaluation of scientific evidence and their specific needs. The purpose for which an animal is kept or exploited is irrelevant – the animal neither knows nor cares. Simply using legislation to support traditional or commercial practices in the absence of evidence of humaneness cannot be justified.

The evidence of the capacity to suffer in vertebrate animals, and in an increasing number of invertebrates (such as crabs, lobsters and octopuses), continues to mount – and yet they are excluded from most legal protection. Are the arbitrary groupings which afford some groups better protection from suffering than others any longer tenable? I don't believe so.

Dividing the animals we exploit into arbitrary groups, each with its own set of rules and practices, perpetuates numerous anomalies and inconsistencies. This is a short list of examples:

- By law any premises used for research using animals needs a premises licence, a named veterinary surgeon and a named animal care and welfare officer. All persons working there need training and certification appropriate to their activity. All activities need to be individually licensed. Official scrutiny is intense. Compare that to premises with farmed animals. Other than for slaughterhouses there are no premises licences, no formal qualifications, no named veterinary surgeon, and no formal recognition of the husbandry practised is required.

- Spring traps for killing stoats need to meet international humane standards, but the spring traps designed to kill moles, rats and mice need not meet any standard.
- The inhumane treatment of a single dog or a horse will dominate the news, leading to calls for stricter sentencing for animal welfare offences, whereas the suffering of perhaps tens of millions of rodents that are killed annually as 'pests' or 'vermin' stimulates little interest.
- It is illegal to dock the tails of cattle and horses but not illegal to dock the tails of sheep.
- Killing of animals in slaughterhouses and research premises is limited to authorised and competent operatives. There is no universal requirement for those killing wildlife, whether it is trapped, poisoned or shot.
- Although the Animal Welfare Act 2006 (and the equivalent Acts in other UK countries) includes fish within the definition of protected animal, there is an exemption for anything that 'occurs in the normal course of fishing' (Section 59 of the Act). There are no exemptions for mammals or birds.
- Some organisations work hard for continuous improvements in animal welfare. Others do not. For example, the BTO strives to reduce the impact of trapping and ringing by collecting, analysing and acting on data. This contrasts with other wildlife interventions: for instance, despite there being ample opportunity to collect data on lethal trapping practised by gamekeepers, there are no field data on the mortality and wounding of stoats.

Animal welfare is not the same as animal rights

Quite early in my career I had to get used to being described as a 'bunny hugger'. I was awarded this epithet when I began to question some of the orthodoxies of animal farming, wildlife 'management' and the way we treat our pets. It was meant as a jibe but I was secretly quite proud of it.

Recently, as the debate around our relationship with animals has become increasingly vitriolic, and as I have become involved in the ethics of animal exploitation, I have been labelled as a supporter of animal rights. This is a shame. Animal rights is a term that should not be banded about glibly, because it has a rather specific meaning. Animal rights are different from animal welfare, although I accept that the distinction between the two can be unclear.

Animal rights theory is strictly deontological. Deontology is a theory that suggests actions are good or bad according to a clear set of rules. Its name comes from the Greek word *deon*, meaning duty. Only actions that obey these rules are deemed to be ethical. Animal rights theory has it that animals, like humans, have rights such as the right not to suffer and the right to life. Hence some animal rights advocates seek the abolition of most if not all of the inter-actions we have with animals. And not just those we eat or perform experiments on. The purest animal rights advocates would have us keeping no pets.

In contrast, **animal welfare** is consequentialist or utilitarian: animals do not have inalienable rights, but animal welfare seeks to minimise overall suffering and death and maximise utility. This makes animal welfare flexible. Some proponents seek radical change up to and including the abolition of most interactions; others accept exploitation provided that the benefits exceed the suffering caused – the harm–benefit analysis described in Chapter 10.

You can easily see where confusion between the two positions arises. For example, I might be of the view that an animal has a 'right' not to be abused. This might be taken as an indication that I am an animal rights advocate – but given that avoiding abuse is effectively enshrined in law, which in turn reflects societal values, then preventing that animal from being abused is simply me discharging my responsibilities as a citizen. Of course, it gets more complicated when we realise that the law allows for many procedures that animal rights activists and many animal welfare advocates would consider inhumane: everything from the confinement and mutilation of farmed animals, to the trapping and poisoning of wildlife, to the use of animals to help determine the safety of vaccines for people.

A view that an animal has a right not to be abused is not incompatible with a position that allows for limited and carefully managed use or exploitation of animals. Determining where the boundary falls between the acceptable and unacceptable use of animals is rarely straightforward. Which is why the animal rights abolitionist position is attractive to many – there need be no backsliding and no compromise.

But those of us committed to a utilitarian position must not be complacent. Merely because tradition dictates it or the law allows it does not mean it is humane or acceptable. Every activity that I describe in this book, every trap, every cage, every husbandry system, every poison and every tag needs scrutiny, which must include a process of periodic review. And when it cannot be demonstrated to be humane and/or there is an unfavourable harm–benefit analysis, then either substantial modification is required or its continued use must be prohibited. It's that simple.

Is sentience important?

I introduced the concept of animal sentience in Chapter 1. Remember, sentience is the capacity of an animal to experience different feelings such as suffering or pleasure. Animals known or believed to be sentient, irrespective of their 'use', should be given societal and legal protection appropriate to their needs. But what does that really mean?

First, getting unequivocal evidence of sentience in each exploited species is an impossible task. As set out in Chapter 1, rather than agonising about a lack of data we must simply give many species the benefit of the doubt. Second, once a legislator recognises a particular species as sentient or, by acting precautionarily, makes the assumption that it is, then the needs of that animal must be met, irrespective of the nature of our interactions or the category in which the species sits. Even allowing for the possibility that there are degrees of sentience, this offers the prospect of resolving the many anomalies and iniquities that beset our relationships with animals.

The European Council and the European Parliament, the primary legislative bodies of the EU, are obliged, following the Treaty of Lisbon, to consider animal sentience when making legislation.[1] Many have argued that the obligation has had little effect, since the 'art of the possible' in politics means that a consideration of sentience when making animal welfare legislation is merely one factor alongside economics, international trade and commercial considerations. However, there is no doubt it is a step forward for a major legislative body to be bound by such an obligation.

The UK has left the EU and is no longer a signatory to the Treaty of Lisbon. This means that, unless the law changes, the government would be under no obligation to take animal sentience into account when legislating. But, following a high profile campaign, an Act of Parliament has recently been passed which enshrines a similar obligation in legilsation.[2]

What difference will it make? Despite the mounting evidence of sentience and of the capacity to suffer in a wide range of animals, and given that the same obligation has existed in the EU for two decades, it is hard to envisage rapid change to animal welfare legislation.

Call me a cynic, but I can't see the Animal Welfare (Sentience) Act removing the exemption for things that happen during 'the normal course of fishing' happening any time soon. Nor do I expect imminent legislation to prohibit the use of demonstrably inhumane practices such as mole trapping, the dehorning of cattle or the breeding of deformed dogs. It will take some time. But it ought not to be business as usual either. There's scope for considerable change, and a grand, welcome gesture like recognising animal sentience ought not to be celebrated by a little tinkering at the edges. There are matters of principle to address.

What would success look like? I can think of a number of outcomes that would indicate that sentience was being taken seriously:

- We would agree that systems of keeping farmed animals which rely on confinement in barren environments, and which inhibit normal patterns of behaviour, are unacceptable.

- We would drop terms like 'vermin', 'pest' and 'problem species'. These pejorative terms put several species into a category where inhumane control, which would not be acceptable for other animals, is the norm. Likewise the use of the term 'game', since the value ascribed to 'game' species is used to justify the inhumane treatment of other species deemed to be a threat to the production of a 'shootable surplus'.
- We would adopt the principle that any intervention with the potential to affect negatively the physical or mental state of an animal must have a positive harm–benefit analysis. This would cover every intervention, from the relatively benign, such as the fitting of identification tags to trapped birds, to the extreme, such as mutilations without pain relief and inhumane methods of controlling wildlife. Currently, formal analysis of this type applies only to research.
- We would expect that operatives involved in any activities where there is a substantial risk to animal welfare would be demonstrably competent, with periodic reassessment as required. This would include anyone involved in the killing of wildlife for any purpose. At present, only those working on research animals and on the slaughter of farmed animals need to demonstrate an ability to handle and care for animals humanely, although there are a number of industry codes which require operatives to meet a standard.
- There would be effective, risk-based independent inspections of all potentially harmful activities that risk affecting animal welfare.
- There would be a system of independent approval of new and substantially modified systems of husbandry of farmed and other animals, putting animal welfare at the forefront of the approval process.
- There would be an obligation to have independent approval for any system, practice or equipment used for the trapping or killing of wildlife and other animals, showing that it is capable of killing or trapping humanely. Snares, most lethal traps, live-capture traps and most poisons currently in use do not meet this criterion. The rationale for any lethal trapping would be thoroughly tested. It would have to be demonstrated that there is an overriding conservation or commercial reason for using it, and that non-lethal alternatives have been tried and have failed.
- We would expect that animals used in sport are treated humanely and subject to effective independent oversight.

This is not an exhaustive list. It is intended to catch the majority of activities where change is needed, but it is deliberately short on detail since broader consultation would be necessary before comprehensive, meaningful proposals could be developed. However, I hope it is obvious that implementing this short list would have a profound impact on the way in which animals are exploited. It might also have a profound impact on the economy – the end of cheap animal protein, for example. We should make these changes carefully, then, using evidence and forging consensus.

A recognition of sentience in the animals we exploit is seen by some as a step towards a recognition that animals have rights. However, it isn't necessarily so. Recognising sentience is a step towards better animal welfare and an opportunity to ensure that animal exploitation, such as we are prepared to tolerate, is evidence-based and humane.

It's not as if there haven't been changes already: the nineteenth-century prohibition on obviously cruel practices such as bull baiting and dog fighting, and the twentieth-century prohibition on the use of pole traps to maim birds of prey, were major steps forward. Laws to require humane killing in slaughterhouses and to require the use of anaesthesia when dehorning cattle were a recognition that even the commonplace was worth regulating better.

Perhaps of greater importance are more recent changes that were implemented because of the impact on the mental state of the animals kept in the system: the prohibition on the use of sow stalls and tethers, and on the use of barren cages for laying hens, came about not because of the physical effects of confinement – although these effects were clear. Both systems were eventually outlawed because of the evidence that permanent confinement and the associated impact on their behavioural needs was causing large numbers of pigs and chickens to suffer. That is quite a turnaround – albeit a turnaround that took several decades to make.

I'd like to ensure not only that this legacy is maintained but that the enlightened attitude of regulators in the last century can be rekindled to effect further improvements in animal welfare.

Becoming active in your own right

Having got this far, you might reasonably ask 'What does all this mean to me?' and 'What can I do?'

While adopting an animal rights agenda or becoming vegan is a rational choice, I am an advocate of neither. This book is aimed squarely at people whose concern and interest has them resting somewhere on the spectrum between quite concerned and mildly outraged about our treatment of animals. How do we take that concern and outrage and make a difference? And how do we decide what exploitation we are prepared to tolerate, particularly as there is limited information available?

While the numbers of vegans and vegetarians continue to increase, and Cartesians are either becoming rare or have simply gone rather quiet, I believe there is a substantial number of people who want to take a more nuanced position, to form opinions and take action based on evidence rather than propaganda and hysterical campaigning. While there's room for all shades of opinion, we should be mindful of facts when forming our views. Intelligent people draw on evidence, not prejudices amplified on social media. To this end, much of this chapter is intended to help you form and refine those views, and to suggest how they might be put into practice.

We've been through the many ways in which we exploit animals. While much exploitation is accepted or at least tolerated by society, either because we don't know what goes on, or because we don't care, it doesn't mean that you should continue to accept all – or any – of it.

Setting a limit to what we tolerate

I've described many types of animal exploitation. So much of it is routine, institutionalised or, by luck or careful design, not visible to the general public. A great deal of it takes place without us having any exposure to what is happening. That includes everything from animal experimentation to wildlife 'management' to intensive poultry production. It goes on at scale and, perversely, the larger the scale, the more we seem to accept it. Annually, in the UK, we rear a billion broiler chickens, a substantial proportion of which develop crippling bone and joint defects before they reach slaughter weight; tens of millions of farmed animals are routinely mutilated – to remove tails, horns and testicles, most of it without anaesthesia or postoperative analgesia; 40 million pheasants are reared for shooting, requiring the killing of countless native birds and mammals using inhumane methods to protect them before they are shot at by people with no mandatory training; millions, perhaps tens of millions, of rodents are killed with poisons known to be markedly inhumane; and almost three million animals are used in research. All of it is legal and a substantial part of it takes place with little or no independent scrutiny.

The scale is mind-boggling. It's hardly surprising that people choose to campaign for a ban on the keeping of a few tens of wild animals in circuses, or for longer sentences for animal welfare offences. You can at least get your head around these things. They are worthy causes, but because of their scale they are hardly the most important animal welfare issues. Given the number of broilers produced every year in Britain, relatively minor beneficial changes in the way they are reared would have a much greater impact on animal welfare overall.

How do you make choices and drive change? It's difficult, and that's primarily because of the lack of transparency. Why not open the doors and let everyone take a look?

Opening the doors

Businesses and institutions such as farms, slaughterhouses and research laboratories are wary of transparency and openness. And not without good reason. Too often they have had their fingers burned and their activities misrepresented in the media, to the extent that the meat industry verges on paranoia when it comes to publicity. And the security around gaining entry to an animal research establishment is greater and more intrusive than the requirements for accessing some Ministry of Defence buildings.

The shooting sector keeps its counsel and actively avoids scrutiny. Shooters have always been rather coy and publicity-shy, but the growing number of people actively campaigning against shooting has reinforced that ethos. Conservation organisations are hardly role models either, often being cagey about what animals they kill to protect the habitats and threatened species they look after.

But things are changing. Farming organisations are now making an effort. Events like Open Farm Sunday (farmsunday.org) help to educate and inform consumers. The shooting sector is actively making a case for its activities – but, in my experience, it needs to do more to demonstrate that the welfare of the animals involved, including those that shooters consider to be 'pests', is taken seriously.

Conservation organisations have been coy about, for instance, culling deer, although some, like the RSPB, have at least made a start.[3] Rather than keeping schtum, one way of winning over your membership and stemming the flow of misinformation from those with an agenda is to explain and inform. Explaining the evidence base, say, for time-limited and geographically restricted killing of crows to protect nesting curlews while the population recovers may not convince the entire membership, but it will help quieten the wilfully misinformed. And it will mean there is clear blue water between conservation NGOs and shooting organisations, who believe that year-round, unfettered and unreported killing of corvids is necessary to protect their interests.

Sections of the meat industry appear to have a bunker mentality. Each time there is a video exposé of the worst excesses of their activities, they remain silent, I guess in the hope that the controversy will simply fade away. Meanwhile, another few hundred or thousand people become vegans and are unlikely to reverse their decision. The industry needs to do more – improving standards and ejecting those that can't or won't apply good practice and the law, for example.

Research establishments are in a difficult position because of the number and types of people that are vehemently and implacably opposed to the use of animals in science. How do they open up while managing the risk of potentially disruptive persons gaining entry? There's no easy answer, but perhaps the way to inform and inspire confidence is not through guided tours around research laboratories but to open them up to scrutiny through other means. I'll return to this later.

It is unfortunate, but the main way many of us get information about these activities is by watching video footage of slaughterhouses, research establishments and farms taken clandestinely by organisations with an abolitionist agenda. Some horrible activities have been uncovered, and sometimes animal welfare offences are prosecuted, but what gets shown is hardly typical. In my experience, people working in these establishments want to do a good job. But in the absence of anything else it is hard for people to form a more favourable view. Rather than opening up, the organisations targeted often drive themselves deeper into the bunker with ever more layers of security. This merely reinforces the notion that they have something to hide. Clandestine filming may not do much for greater openness – although it could be argued that the introduction of mandatory closed-circuit television in slaughterhouses was stimulated by the exposés.

People with no direct involvement in animals need a say. Society as a whole, despite having a stake in how animals are exploited, has largely been excluded from decision making. And that breeds mistrust and suspicion. Why would or should anyone have confidence in decisions where representation from large sectors of society is absent? More inclusive decision making is not without risk. Decisions may take longer, and gaining consensus may prove to be impossible. And I have no doubt that many of the more obscure practitioners of animal exploitation will be nervous about the notion. But that doesn't mean it shouldn't be done.

There are a number of ways to make a difference. Some of them are open to us all as individuals, and some depend on us acting collectively, as a society. And there are other, as yet untried ways that could and should influence animal exploitation. But first let's take a look at you, acting as an individual.

The consumer and the citizen

Before delving into the detail, we need to consider the difference between two sorts of individual, the consumer and the citizen. A consumer is a person who buys goods or services for his or her own use. The definition of citizen is more fluid, and the word means much more than its historical derivation ('a resident of a city'). A more modern definition is 'a resident of any particular place to which the subject feels he or she belongs'. By convention, that residence confers upon the citizen rights (such as the right to vote) and responsibilities (paying taxes, for example).

Although there is considerable overlap between the consumer and the citizen, there are subtle but important differences in the way each of us, depending on which of the two roles we are currently in, can influence events. When the consumer purchases goods and services, for instance, as a citizen that person is obliged to pay VAT. But consumer and citizen are not the same. This is particularly important when it comes to animal exploitation.

Exerting influence as a consumer

Consumers, to use that hackneyed phrase, can vote with their feet. Put simply, you can avoid the excesses of dog breeding, the cruelty of the fur trade, and the poor practices of intensive pork and poultry production by not buying the products on offer. Instead, take veterinary advice and end up buying a dog-shaped dog rather than a deformed one with a lifetime of health problems to deal with. Or buy a coat lined with man-made insulation rather than a fur coat. Or buy a slower-grown oven-ready chicken rather than a dopey, misshapen broiler. Or, alternatively, don't have a dog and don't eat chicken. You might still need that coat, however.

In choosing what to buy, consumers can affect the sales figures of a particular product or service. It explains the near collapse of the fur trade in Britain – because of public concerns about the welfare of trapped wild animals and animals farmed for their fur, it has become socially unacceptable to wear fur and now hardly anyone buys it.[4] Consumers appear to be applying the same tactics to food. Supermarket chains know this. They watch trends very carefully and act quickly to change their offer accordingly. Which is why they are extending their vegan and vegetarian ranges, and why most have made the decision no longer to stock eggs from caged hens. But while I want people to make informed choices, we shouldn't be deluded into thinking that one person or even 50 people avoiding broiler meat is going to precipitate wholesale reform of the poultry industry. However, by acting collectively to avoid a particular product, perhaps in response to a campaign, the consumers can effect change. That is why the BVA and other organisations are actively campaigning against brachycephalic dogs, in the hope that if consumers avoid buying the puppies, breeders will start to breed from dogs with better conformation.

Exerting influence as a citizen

There are whole areas of animal exploitation where consumer action is not and will not be effective. That is because there is no 'product' to buy – or avoid. Suppose you object to wildlife 'management' because you feel, like me, that the trapping and snaring of foxes and stoats is inhumane, and that, despite the claimed benefits – that killing foxes and stoats protects pheasants and grouse – these do not override the harms. You might argue that avoiding eating 'game' birds would have a bearing on the commercial health of shooting. It might, but I doubt it. Your boycott will have little impact, given that a substantial number of the carcasses of shot pheasants are simply discarded and never find their way to a retail outlet. Similarly, if you had similar concerns about deer and stopped eating venison, it would have little impact on the numbers of deer shot, or on how they are shot, since so many are killed to protect forestry and crops.

By the same token, it doesn't take a great deal of thought to realise that your influence, as a citizen (or indeed as a consumer), over the use of animals in research is likely to be low. Sure, campaigning has forced the regulator to reduce the numbers of animals used, and perhaps consumer pressure contributed to the ban on the use of animals in testing cosmetics. But this is small beer. Real change comes from influence on the inside, and later in this chapter I'll explain how this is happening.

For those issues where consumer boycotts are unlikely to be effective, you might be able to play more of a role as a citizen, particularly if you act collectively. As a citizen you can seek to influence policy by responding to government consultations, becoming active in conservation and animal welfare organisations, writing to your Member of Parliament, and joining demonstrations. Many of my more cynical colleagues and friends involved in animal welfare and conservation believe that none of that will make much difference. But that's a bit of an exaggeration – some campaigning leads to real change.

To be successful, those seeking change have to campaign, and campaign long and hard. Carefully formulated evidence-based campaigns which engage decision makers, opinion formers and politicians can work. But they take time, money and commitment. And don't think that simply writing form letters will be sufficient. Believe me – bombarding politicians and regulators with hundreds of identical postcards each making exactly the same point is a waste of time. They get counted, piled in a corner of the office and then promptly forgotten. To be influential the campaign needs a clear objective supported by a robust evidence base and backed by a network of like-minded and committed people. There are two good examples of successful, well-organised campaigns which are changing animal welfare for the better. The Better Chicken Commitment, discussed in Chapter 5, is encouraging (or is it shaming?) retailers to make improvements in the way that animals are treated.[5] Crustacean Compassion has successfully campaigned to have the Animal Welfare Act amended so that crustaceans such as lobsters and crabs enjoy the same protection as vertebrates.[6] The timescale of both campaigns is measured in years, not weeks or months. Note that the campaigns sought to influence two different constituencies – for good reason the Better Chicken Commitment is directed at retailers, while Crustacean Compassion is directed at regulators. There is a particular type of animal exploitation that seems largely refractory to these tactics, however, and that is wildlife management.

Why is it so difficult to influence wildlife management?

Wildlife law is skewed generally in favour of landowners, and the socioeconomic benefits of wildlife are rarely if ever taken into account when developing policy. This means that the likelihood of animal welfare arguments holding sway on their own is very low. Meanwhile, conflicts over the 'management' of wildlife grind on, with campaigns against the use of Larsen and similar traps, snares and spring traps alongside perennials such as badger killing and fox hunting. They are likely to make only slow progress while the odds are so heavily stacked against the citizen.

The various ways in which wildlife is exploited for sport, 'pest' control or even conservation are inconsistently regulated. There is little independent scrutiny. There are many examples – the use of cage traps to catch crows and other birds, the licensed killing of badgers in an attempt to control bovine tuberculosis, and the killing of moles using unregulated and untested traps. Quite apart from the questionable humaneness of each of these, there is the question of 'why?'

However much you find these activities objectionable, and despite the killing taking place right over the fence from where you live, involving animals that may even come to your garden, you have little or no influence on them. It is true, for instance, that the principle of licensed badger killing was the subject of extensive consultation – but as I point out in Chapter 6, when the killing starts, local people are neither informed nor consulted unless they have a substantial land holding.

The primary reason for the intractable nature of so many of these disputes is not because of lack of evidence or lack of resolve. It is because the law and most government policy treats animals as property and without an intrinsic value in their own right. Pets, farmed animals and research animals are chattels. That is, they are owned by an individual, a commercial organisation or an institution, and provided the animal is treated within the bounds of the law (or you don't get found out) you have complete control over its experience and ultimate fate.

Wildlife in the UK is viewed differently. Unlike in the United States, where it is considered a public resource, independent of the land or water where it lives, under UK law 'wildlife' doesn't belong to anyone but, again, provided you act within the law (or don't get found out), you can do what you like with it. And once it's dead and it's on your land, it's yours. In the UK, a person cannot own, 'absolutely', a wild animal while it is alive. However, in certain situations a person may be regarded as being the 'qualified' owner of a live wild animal. Where ownership is 'qualified' the person does not have an exclusive or permanent right to the wild animal.

There we have it. The only value society places on animals is a monetary one associated with ownership while it is alive and, in the case of wildlife, once it is dead on your land. Of course, ownership confers upon the owner (or someone else in charge for the time being) certain obligations about care and protection from harm. But society doesn't generally have a say in how this is done, or any direct responsibility, except via the arcane, slow and generally unsatisfactory process of consultation about government regulation.

This inability of the citizen to influence the way in which wildlife is treated in Britain explains in no small way the success of Wild Justice, an organisation described by its founders, Chris Packham, Mark Avery and Ruth Tingay, as having been 'set up to fight for wildlife'. They go on to say that wild animals 'can't take legal cases in their own names but, with your help, we will stand up for wildlife using the legal system and seeking changes to existing laws'.[7]

Not only has Wild Justice been successful in effecting changes in wildlife protection law through legal challenge, it has also captured the interest and imagination of many citizens concerned about the way our wildlife is treated. Legal challenges have been funded almost entirely by donations from large numbers of individual citizens. The total amounts involved are not trivial, reflecting both the concerns of citizens and their frustration at a lack of influence through other means.

While I applaud the efforts and success of Wild Justice, I find it profoundly dispiriting that the only apparently successful means of challenging existing wildlife law and its interpretation is by recourse to the courts. There has to be a better way.

Towards evidence-based and inclusive decision making

Whether we like it or not, most improvements in the welfare of the animals we exploit, other than through the exercise of consumer choice, come about through the development and implementation of government policy – generally through legislation. In my experience, governments set out to develop new policy in a structured way but 'bumps in the road' affect the direction of travel. Proposals become diluted or sometimes simply abandoned by dint of confounding factors such as lack of evidence, burgeoning cost estimates, practical considerations and competing political pressures.

During this process there is consultation with interested parties. Depending on the nature of the proposal, the consultation might be 'broad brush', where the questions asked deal with the overarching principles of the legislation. For example, in making animal welfare law, questions posed might include 'What should be the scope of the proposed legislation?' or 'Should it cover all animals, or just mammals and birds?' Respondents, as well as answering the question, are asked to provide a rationale backed with evidence.

In other consultations, when the matter is primarily technical, the consultation might be more limited, for instance 'What are the most important features in a furnished cage for a laying hen?' or 'What is the minimum age that a sheep should be given analgesia during tail docking?' Again, respondents are expected to provide supporting evidence. What is lacking in these hypothetical questions is the opportunity to ask whether a laying hen ought ever to be confined in a cage, or whether a sheep's tail should ever be cut off.

In practice, the individual citizen has very limited opportunity to influence the ethics of animal exploitation. The ban on unfurnished cages for laying hens is an exception, but that took over 20 years to achieve. And even then, substantial concerns remain about the welfare of laying hens husbanded at scale – whatever system is used.

Legislation based on ethical principles has been enacted. but the effects are generally small. The ban on fur farming, introduced on ethical grounds in 2000, at the time of enactment applied to fewer than 10 premises across the UK.[8] When the ban on keeping and using wild animals in circuses was enacted in 2019 there was a grand total of two circuses with wild animals, boasting between them six reindeer, four zebra, three camels, three raccoons, a fox, a macaw and a zebu.[9] It is easier to apply ethical principles when the number of businesses affected and the number of animals involved are small. Of course, other similar businesses may have got out as they saw the writing on the wall in the months before the law was made. And both Acts prevent new fur farms and new circuses with wild animals from being set up.

Nevertheless, it is hard not to conclude that this is relatively low-impact, crowd-pleasing legislation that deflects attention from discussing the ethics of the farmed animal and wildlife exploitation that takes place daily on a massive scale – for example, since 2013, tens of thousands of badgers killed annually in circumstances where the government has failed to demonstrate that it is done humanely.

Badger killing is an issue where, despite legitimate conservation and animal welfare concerns, the citizen is effectively cut out from any influence over policy and practice. Successive opinion polls show the majority of the public is against licensed badger killing, but their views are ignored. Of course, most of us have no financial stake in cattle production, the protection of which is the reason that badgers are killed. The first tuberculous badgers were identified in 1973 and the official killing of badgers began in 1975.

Since then there have been six independent reviews of the science behind the government's strategy to eradicate bovine tuberculosis. At no point has a government-commissioned review considered any ethical questions concerning the systematic killing of native wildlife. Nor has any review considered the socioeconomic impacts of the removal of large numbers of native wildlife. Constrained by their terms of reference, the reviews have concentrated on the disease and its economic impacts.

Public resistance to badger killing (and fox hunting, broiler farming, research involving animals, the trapping and poisoning of 'pests', and zero-grazing of dairy cattle) will continue for as long as the animals involved are considered primarily as chattels with no social value, and for as long as ethical concerns remain wholly subordinate to financial considerations. There has to be a better way.

Gaining consensus

Decision making can be improved where there is a will. We need systems which, along with financial, economic and technical considerations, are capable of placing ethics, evidence and people at the centre of the process. A number of ideas developed by conservationists, academics and animal welfare advocates have begun to emerge. Used properly, these approaches offer the prospect of real change.

A pioneering approach developed by an international team of experts proposes using a set of consensus principles for ethical wildlife control.[10] It consists of seven principles which includes altering the human practices that cause human–wildlife conflict, minimising animal welfare harms and considering community values as well as scientific, technical and practical information. That these principles have been cited numerous times in publications concerning wildlife as diverse as dolphins, beavers, coyotes, cottontail rabbits, vampire bats and feral pigs is testament to the practical value of such an approach.

Adopting and applying a set of principles will not automatically solve the problem. Gathering evidence and involving more people in decision making will inevitably slow down the process. But, provided the process is followed carefully, there is greater likelihood that lasting consensus will emerge – in contrast to the current exclusionary and opaque processes.

Applying these principles will not necessarily provide a repeatable outcome, since the views and values of those involved will have a bearing on the outcome. For example, two papers using the same consensus principles, published in the same year and on the same subject, reached very different conclusions about the licensed killing of badgers to control bovine tuberculosis.[11] Both analyses suffered from engaging too few people; it is important that the process involves a variety of disciplines and stakeholders to ensure that a robust consensus emerges.

Assessing the ethics of animal exploitation

Currently, with few exceptions, restrictions on the development and commercialisation of new ways of exploiting animals are minimal. Nor is there any effective review of the myriad ways of exploiting animals that have developed over the last few hundred years. This means, for instance, that electric-shock collars for training dogs can be placed on the market without any need to demonstrate that they are humane. Use of these 'training aids' is prohibited in Scotland and Wales and there are plans for a similar ban in England. A system of pre-approval would have likely prevented these from being offered for sale in the first place. Similarly, new systems of farmed animal husbandry can be adopted without a clear understanding of the impacts on behaviour and welfare – for example, zero-grazing systems for dairy cows. In both cases, government advisors and legislators are left scrabbling for evidence and the means, should it be necessary, to either endorse or curtail the system.

Worse is our repeated failure to address common and established practice where we already have evidence that it is inhumane or where intuitively it appears inhumane and there is no objective evidence to the contrary. Examples of these include the routine mutilation of farmed animals and the use of wire snares to trap rabbits and foxes.

I believe that the burden of proof should rest on the proponents of a particular method of animal exploitation, new or old. There should be a requirement to demonstrate, at the development stage and prior to widespread adoption, that any new system of keeping cattle or training dogs, for instance, is humane. Had such a principle been in existence at the time when systems for caged laying hens and sow crates were first developed, it is quite possible that they might not have been approved for widespread use. This idea is gaining currency, and a uniform national testing and approval procedure for animal husbandry systems has been proposed for use in Germany, though it has yet to be adopted.[12]

The same principle could be applied to established methods of killing wildlife. For example, there could be a period of, say, five years to provide evidence that snares are humane. If no objective evidence emerged then their subsequent use would be prohibited. One way of amassing evidence about the humaneness of snares would be to collect data on their impact on the behaviour of trapped animals, any injuries caused, and physiological indicators of stress. Such a trial would probably need a licence under ASPA, but an application would most likely be rejected on ethical grounds. If evidence were needed about the huge disparity between what we accept (or are obliged to tolerate) in the treatment of animals in different circumstances, then this is surely it.

The Animal Welfare Impact Assessment (AWIA), a proposal by academics Steven McCulloch and Michael Reiss, offers the prospect of a structured and objective method of assessing impacts on animal welfare where substantial changes to policy are proposed.[13] They argue that an AWIA should be completed for policy options that have the potential to significantly impact the interests of sentient animals. The process is similar to other routine formal assessments that are required when major projects are planned, covering social, economic, environmental and health impacts, each of which is a well-established mechanism used in their respective fields. The assessment covers all the harms and benefits for species affected by policy options under review.

In a similar vein, McCulloch and Reiss have called for an 'Ethics Council' to inform the UK government on policy that significantly impacts sentient species. They argue that such a body is necessary for 'just and democratic policy making in all societies that use sentient nonhuman species'.[14] It is worth noting that this is broadly similar to the government's expert Animal Sentience Committee. The committee is expected to identify relevant policies and advise ministers how government decisions might take into account animal sentience. At the time of writing, the government was recruiting members.

On the face of it, these changes would be simple, logical and easy to implement. In practice, it will be complicated and difficult. Libertarian and conservative attitudes and the proponents of the small state will object because of the perceived intrusive nature of the process and the associated bureaucracy. There is also the real risk that states acting unilaterally to outlaw certain practices might simply 'offshore' the problem: for example, if country A prohibits a certain procedure for a particular species of farmed animal, country B might take advantage of the lower costs it enjoys by continuing with the practice and 'dump' surplus cheap animal products on country A's markets. This argument is frequently cited by Britain's farmers worried that they will be undercut as the government agrees trade agreements with countries believed to have lower farmed animal welfare standards than Britain. At the other end of the spectrum, vegans and animal rights advocates will argue that evidence-based reviews and change are simply a mechanism for perpetuating existing practice with minor modifications, when in fact all exploitation should cease. I'm not expecting it to be easy, but with the necessary political will, and with good governance that includes citizen engagement, there is every reason to believe that change can be effected.

A citizens' assembly

Systems that objectively assess the animal welfare impact of emerging technology have the potential to drive, or at least influence, change. And a consideration of the ethics of any methods of animal exploitation, along with the evidence, and drawing on a variety of disciplines, is an important first step for change. Where does that leave the concerned and marginalised citizen? One way of gaining a broader consensus is to involve more people, ensuring that diverse interests and backgrounds are represented.

A citizens' assembly is a body formed to deliberate on important issues. The people who take part are chosen so that, as far as possible, they reflect the wider population. Citizens' assemblies give members of the public the time and opportunity to learn about and discuss a topic, before reaching conclusions. Assembly members are asked to make trade-offs and arrive at workable recommendations. Assemblies have been used to deliberate on climate change, abortion and same sex marriage, among other subjects.

Although citizens' assemblies have been credited with successes in solving particularly divisive issues, they have come in for criticism. Their legitimacy rests on the idea that a small group of citizens randomly selected to reflect the age, education level, wealth and gender make-up of the general population does indeed represent the public as a whole. That is not always clear-cut. The time involved in meetings and background reading is considerable and may exclude people with heavy work commitments and care responsibilities. Travel and accommodation costs can make the process expensive. Protracted deliberations might lead to a high dropout rate. Despite these potential drawbacks, interest is high and there seems every reason to adopt a similar process to resolve contentious animal exploitation issues, particularly if the process is built around a formal ethical framework.

Public engagement in research

Research is done in the name of and for the benefit of society. It is largely paid for out of taxation or by charitable donations. Either way, we are paying for it. Legislation that governs the use of animals in science is made in our name by parliamentarians. It is only right that citizens have some influence over how it is done. As we saw in Chapter 10, it is a legal requirement for each licensed establishment to run an Animal Welfare and Ethics Review Body (AWERB). This strikes me as a reasonable alternative to the extremes of complete transparency (with the attendant security risks) and working completely out of sight.

Although an AWERB must be constituted with scientists and people with statutory animal welfare responsibilities, there is an expectation in official guidance that AWERBs will 'actively seek a wider membership taking into account, in a transparent manner, the views of people who do not have responsibilities under ASPA, as well as one or more persons who are independent of the establishment'.[15] In practice, 'lay people' are generally appointed from within the organisation or, if from outside the organisation, are carefully vetted.

In an effort to make the AWERB more representative of society, some AWERBs actively seek lay members with no links to the institution. Leicester University, for instance, seeks to widen its AWERB membership in order to take into account the views of lay people who have no personal responsibilities or other involvement in the laboratory use of animals. The university believes that such members provide a different perspective. They view established practice and accepted norms with a 'fresh eye' and bring a societal perspective to the consideration of animal experiments.[16] This is an excellent initiative and meets the need for citizen engagement perfectly.

What happens when you open the door

Whether by accident or design, most animal exploitation takes place out of sight. But is it out of mind too? Some people choose to look the other way, but that is not the most prevalent attitude. When people were surveyed by the polling company Ipsos MORI in 2013 and asked about research using animals, the most consistent response was that they wanted to know more, not less, and that the animal research sector should be more open.[17]

For example, those surveyed wanted clear and detailed information about how many animals are actually used, for what purposes and under what circumstances, how much suffering is caused, and how and why they are killed. Where possible, they wanted the public allowed into research establishments, specifically the laboratories where experiments are undertaken, in order to learn more.

These findings stimulated the research community to produce a concordat with a commitment to greater openness, including detailed information on when, how and why animals are used in research, enhancing communications with the media and the public, and providing opportunities for the public to find out about research using animals.[18] There is now a dedicated website (concordatopenness.org.uk), with annual reporting and a growing number of member organisations in the UK and elsewhere. Security around the establishments remains tight, but the amount of detailed information grows annually, covering everything from the numbers of animals involved to the types of procedures employed, including video footage of the actual procedures.[19]

This process is in sharp contrast to the woeful lack of public accountability of other sectors involved in animal exploitation. While the majority of people may have little or no experience with animals other than what they gain from their pets, the assumption that people are not interested or, worse, that only fanatics and extremists seek change, must be wrong. There is nothing to stop other sectors following the example set by animal research.

Concluding remarks

The exploitation of animals presents society with an ethical problem. We exploit, for our own purposes and gratification, billions of sentient animals, each of which is capable of suffering. We have a moral duty to do this as well as we can. Or stop.

Our society exploits animals in a variety of ways, and we enjoy their companionship and their products. We benefit from them as spectacle, in sport and in science. Some exploitation is relatively benign, some is tolerable with appropriate protections, and some is simply objectionable. Unless we adopt an abolitionist/animal rights position, our animal exploitation must have utility. That is, when harm–benefit is assessed against an evidence base, the exploitation must add some value and be acceptable to wider society.

Despite philosophical differences, competing pressures and the influence of vested interests and tradition, most developed societies, including the UK, have developed a set of principles which govern how we exploit our animals via legislation and practices. These principles continue to evolve, partly in response to better information and partly in response to wider changes in society. But we apply the principles inconsistently, so that the nature and extent of the protection we provide varies widely from species to species and from circumstance to circumstance.

Arbitrary and outdated categories of animals are used to maintain and justify anomalies and inconsistencies which allow for inhumane practices for one species in one circumstance but not in another – despite the fact that most of the animals we exploit are considered sentient and capable of suffering. As I have said before, a rat is a rat is a rat regardless of where and how it is exploited.

The proportion of the population who care is huge. Governments know this. But governments also know that gestures and quick fixes of minor problems make for good publicity and distract attention away from the bigger issues. So we get a ban on wild animals in circuses and longer prison sentences for animal welfare offences. But there is very little discussion about lame dairy cows, deformed broilers, rodent poisons or the use of snares. Too many people have vested interests, and change is either too expensive or likely to put our domestic sector at a commercial disadvantage. It's far easier to expend political capital on populist measures like banning the import of animal trophies by hunters.

We need to reset our relationships with the animals we exploit. We should start by dropping out-of-date categories and developing care and protection based on evidence and values to deliver sensible rules, codes and legislation. Then, by adopting a utilitarian approach to each and every circumstance where we interact with animals, wild or domestic, farmed or research, pets or in sport, we can make decisions about the nature and extent of animal exploitation that we can stand behind. In essence, to drive policy we should adopt, in every case, a form of harm–benefit analysis similar to that applied to the use of animals in science.

The validity and acceptability of new policies will be enhanced when wider society has a greater say than it has at present. This will require people like you – people who have persevered with this book – to demand a greater say. Sure, as a consumer, you can vote with your feet and boycott certain goods, but real change comes when citizens act together and exert influence collectively. We can achieve better animal welfare, but only by becoming engaged.

There are potential risks with change, particularly if the resultant better animal welfare increases costs. Farmers' profits might be hit and, where there is free trade, there is a risk that domestically produced food could be displaced by cheaper and less welfare-friendly options from overseas. One advantage conferred upon EU member states is the consistent nature of welfare standards and the commitment to inspection. It remains to be seen whether British standards will rise or fall now that the UK has left the EU, and whether free trade will confer advantages to consumers and producers or quite the reverse.

I believe that compassionate and empathetic people should yearn for a future where animals are recognised for their sentience, so that animals are only exploited in ways that ensure their needs are met in full *and* where society has concluded from the evidence that the benefits of exploitation outweigh any harms. For this to happen, the doors have to be opened and a light shone into the very depths of every type of exploitation. This will happen only once each sector commits to openness and transparency, and by putting greater trust in the citizen. The commitment of the animal research community to greater openness, stimulated in part by public opinion, is a start, but others need to follow this example. The irony is that the sector considered by many to be the worst of the lot could in fact be the exemplar that drives the rest into treating the citizenry like adults and giving us more say about the lives of the animals we exploit.

Notes

Preface

1 'Suicide rate for vets four times national average': https://www.bbc.co.uk/news/av/uk
-england-somerset-47283443

Chapter 1

1 According to surveys conducted by the Vegan Society, the number of people
claiming to be vegan in Great Britain is estimated at 600,000 (2019) a fourfold
increase since 2014 (www.vegansociety.com). Consumption of meat in Great
Britain, however, has not changed substantially. While both pork and lamb
consumption per capita show a slow but steady decline and beef consumption
remains largely static, the consumption of poultry continues to rise steadily. OECD
(2021) Meat consumption (indicator). https://doi.org/10.1787/fa290fd0-en

2 Bar-On, Y.M., Phillips, R. and Milo, R. (2018) The biomass distribution on Earth.
Proceedings of the National Academy of Sciences 115: 6506–11. https://doi.org/10
.1073/pnas.1711842115

3 HM Government (2021) Monthly statistics on the activity of UK hatcheries and
UK poultry slaughterhouses (data for January 2021). www.gov.uk/government/
statistics/poultry-and-poultry-meat-statistics/monthly-statistics-on-the-activity-of
-uk-hatcheries-and-uk-poultry-slaughterhouses-data-for-january-2021

4 Statista (2019) Import volume of poultry meat in the United Kingdom (UK) from
2003 to 2019. www.statista.com/statistics/316405/poultrymeat-import-volume-in
-the-united-kingdom-uk

5 Compassion in World Farming (2019) Statistics: Broiler Chickens Report.
www.ciwf.org.uk/media/5235303/Statistics-Broiler-chickens.pdf

6 Statista (2021) Chicken meat production worldwide from 2012 to 2020. www.statista
.com/statistics/237637/production-of-poultry-meat-worldwide-since-1990

7 Aebischer, N.J. (2019) Fifty-year trends in UK hunting bags of birds and mammals,
and calibrated estimation of national bag size, using GWCT's National Gamebag
Census. *European Journal of Wildlife Research* 65: 64. https://doi.org/10.1007/
s10344-019-1299-x

8 Mason, G.M. and Littin, K.E. (2003) The humaneness of rodent pest control.
Animal Welfare 12: 1–37.

213

9 National Statistics (2021) Annual statistics of scientific procedures on living animals, Great Britain 2020. Presented to Parliament pursuant to section 21(7) and 21A(1) of the Animals (Scientific Procedures) Act 1986. Ordered by the House of Commons to be printed 15 July 2021. https://assets.publishing.service.gov .uk/government/uploads/system/uploads/attachment_data/file/1002895/annual -statistics-scientific-procedures-living-animals-2020.pdf

10 EU legislation, first enacted in 2009 and subsequently amended numerous times, has brought about a prohibition on the use of animals in the development of cosmetics, ingredients and the associated safety testing. Regulation (EC) No 1223/2009 of the European Parliament and of the Council of 30 November 2009 on cosmetic products. https://eur-lex.europa.eu/legal-content/en/ALL/?uri=CELEX %3A32009R1223

11 de Waal, F. (2016) *Are We Smart Enough to Know How Smart Animals Are?* Granta, London.

12 IASP Task Force on Taxonomy (1994) Part III: Pain terms, a current list with definitions and notes on usage. Classification of chronic pain. www.iasp-pain.org/ resources/terminology

13 Dawkins, M.S. (2017) Animal welfare with and without consciousness. *Journal of Zoology* 301: 1–10. https://doi.org/10.1111/jzo.12434

14 Sneddon, L.U., Elwood, R.W., Adamo, S.A. and Leach, M.C. (2014) Defining and assessing animal pain. *Animal Behaviour* 97: 201–12.

15 (a) Animal Welfare Act 2006. Section 1.1 limits the scope of the Act to vertebrates. www.legislation.gov.uk/ukpga/2006/45/contents. (b) Animals (Scientific Procedures) Act 1986 Amendment Regulations 2012. Regulation 1.1 amends the primary Act to extend the definition of protected animal to any living cephalopod. www.legislation.gov.uk/uksi/2012/3039

16 The Animal Welfare (Sentience) Act 2022 https://www.legislation.gov.uk/ ukpga/2022/22/enacted

17 Singer, P. (1975) *Animal Liberation: Toward an End to Man's Inhumanity to Animals.* Cape, London.

18 Singer, P. (2011) *Practical Ethics*, 3rd edition. Cambridge University Press, Cambridge.

19 The Red Tractor scheme (redtractor.org.uk) is the UK's largest food standards scheme, with 46,000 farming members. It is a not-for-profit company which seeks to ensure that food is traceable and safe. The RSPCA Assured scheme (www .rspcaassured.org.uk) has over 3,000 farm and food production members. A team of assessors and RSPCA farm livestock officers check farm animals have been well cared for to strict RSPCA welfare standards. The welfare standards are claimed to cover the whole of an animal's life, from their health and diet to environment and care.

20 In March 2016, a poll of 1,009 Scottish adults conducted by YouGov on behalf of the More For Scotland's Animals coalition found that 76% supported a ban on the sale and use of snares in Scotland. www.onekind.scot/76-of-scots-want-to-see-a-ban -on-the-use-of-snares-why

21 Schedule 2.c of the Animal (Scientific Procedures) Act 1986, as amended, requires that each premises licence holder must include a condition requiring the holder to establish and maintain an Animal Welfare and Ethical Review Body (AWERB). www.gov.uk/government/publications/animal-welfare-and-ethical-review-bodies -awerb-subgroup-terms-of-reference

Chapter 2

1   Milman, O. (2016) 'Ethical down': is the lining of your winter coat nothing but fluff? *The Guardian*, 14 January 2016. www.theguardian.com/world/2016/jan/14/winter-coat-ethically-produced-down-goose-feathers

2   Sections 40 and 41 of the Animal Health Act 1981 impose a minimum value on a horse or pony intended for export. This is intended to stop any trade in low-value horses and ponies for slaughter. Periodically the values are increased with inflation. These rules were carried over from the Diseases of Animals Act 1950 to the 1981 Act and hence pre-date the UK's entry to the EU. Such a hindrance on trade was anathema to the EU but a blind eye was turned. Attempting to overturn the measure would have expended too much energy and political capital.

3   League Against Cruel Sports (2015) *Vets Against Hunting*. LACS, Godalming. www.league.org.uk/media/filer_public/56/b0/56b06a04-8eae-407c-b397-e406b0b7231f/vets_against_hunting_report_2015.pdf

4   Defra (2007) *Protecting Whales: a Global Responsibility*. https://assets.publishing.service.gov.uk/government/uploads/system/uploads/attachment_data/file/183344/protecting-whales__1_.pdf

5   Cooke, J.G. (2018) *Balaenoptera acutorostrata*. The IUCN Red List of Threatened Species 2018: e.T2474A50348265. https://doi.org/10.2305/IUCN.UK.2018-2.RLTS.T2474A50348265.en

6   Whale and Dolphin Conservation Society and Humane Society of the United States (undated) *Dead or Alive? A Report on the Cruelty of Whaling*. www.humanesociety.org/sites/default/files/archive/assets/pdfs/whaling_cruelty_report.pdf

7   Mech, L.D. and Boitani, L. (2003) *Wolves: Behaviour, Ecology and Conservation*. University of Chicago Press, Chicago and London.

8   Consorte-McCrea, A., Bainbridge, A., Fernandez, A. *et al.* (2016) Understanding attitudes towards native wildlife and biodiversity in the UK: the role of zoos. World Sustainability Series. https://doi.org/10.1007/978-3-319-47883-8_17

9   Environment, Food and Rural Affairs Committee (2019) Controlling dangerous dogs. Ninth Report of Session 2017–19. https://publications.parliament.uk/pa/cm201719/cmselect/cmenvfru/1040/1040.pdf

10 Tulloch, J.S.P., Owczarczak-Garstecka, S.C., Fleming, K.M. *et al.* (2021) English hospital episode data analysis (1998–2018) reveal that the rise in dog bite hospital admissions is driven by adult cases. *Scientific Reports* 11: 1767. https://doi.org/10.1038/s41598-021-81527-7

11 Scottish Government (2019) Sheep attacks and harassment – research: www.gov.scot/publications/sheep-attacks-harassment-research/pages/1/

12 Morris, S. (2018) National Trust criticised after hiring marksman to cull wild boar. *The Guardian*, 19 November 2018. www.theguardian.com/environment/2018/nov/19/national-trust-criticised-after-hiring-marksman-to-cull-wild-boar-stourhead-estate

13 Mayer, J.J. (2013) Wild pig attacks on humans. *Wildlife Damage Management Conferences: Proceedings* 151: 17–35. http://digitalcommons.unl.edu/icwdm_wdmconfproc/151

14 UK Squirrel Accord (undated) Fertility control research: https://squirrelaccord.uk/squirrels/fertility_control

15 Ritchie, J. (1920) *The Influence of Man on Animal Life in Scotland: A Study in Faunal Evolution*. Cambridge University Press, Cambridge.

16 Hale, M.L., Lurz, P.W. and Wolff, K (2004) Patterns of genetic diversity in the red squirrel (*Sciurus vulgaris* L.): Footprints of biogeographic history and artificial introductions. *Conservation Genetics* 5: 167–79. https://doi.org/10.1023/B:COGE .0000030001.86288.12

17 Harrison, R. (2013) *Animal Machines: The New Factory Farming Industry*, revised edition. CABI Publishing, Wallingford.

18 Brambell, R. (1965) *Report of the Technical Committee to Enquire into the Welfare of Animals Kept Under Intensive Livestock Husbandry Systems*, Cmd. (Great Britain. Parliament), HM Stationery Office.

19 Farm Animal Welfare Council (undated) The Five Freedoms. https://webarchive .nationalarchives.gov.uk/20121010012427/http://www.fawc.org.uk/freedoms.htm

20 For example, Council Directive 2008/120/EC lays down minimum standards for the protection of pigs, including space allowances for certain categories of growing pig. https://ec.europa.eu/food/animals/animal-welfare/animal-welfare-practice/ animal-welfare-farm/pigs_en

21 Directive 1999/74/EC on the minimum standards for keeping egg-laying hens banned conventional battery cages in the EU from 1 January 2012 after a 13-year phase-out. Colony cages can be used, but these must be furnished with nest boxes, dust baths and perches.

22 The recently enacted Animal Welfare (Sentience) Act 2022 includes decapods and cephalaopds within its scope. This provides the means for Parliament to protect these species in a way similar to mammals and birds. The Animal Welfare (Sentience) Act 2022 https://www.legislation.gov.uk/ukpga/2022/22/ enacted

23 Mellor, D.J., Beausoleil, N.J., Littlewood, K.E. *et al.* (2020) The 2020 Five Domains Model: including human–animal interactions in assessments of animal welfare. *Animals* 10: 1870. https://doi.org/10.3390/ani10101870

24 Mutilations (Permitted Procedures) (England) Regulations 2007. Statutory Instrument no. 1100 of 2007. www.legislation.gov.uk/uksi/2007/1100/contents

25 Defra (2021) Structure of the agricultural industry in England and the UK at 1 June. www.gov.uk/government/statistical-data-sets/structure-of-the-agricultural -industry-in-england-and-the-uk-at-june

26 Home Office (2019) Animals in Science Regulation Unit: Annual Report 2018. https://assets.publishing.service.gov.uk/government/uploads/system/uploads/ attachment_data/file/887289/Animals_in_Science_Regulation_Unit_annual_report _2018.pdf

Chapter 3

1 Defra (2021) Total Income from Farming in the United Kingdom, second estimate for 2020. https://assets.publishing.service.gov.uk/government/uploads /system/uploads/attachment_data/file/1049674/agricaccounts_tiffstatsnotice -16dec21i.pdf

2 Clark, B., Stewart, G.B., Panzone, L.A. *et al.* (2016) A systematic review of public attitudes, perceptions and behaviours towards production diseases associated with farm animal welfare. *Journal of Agricultural and Environmental Ethics* 29: 455–78. https://doi.org/10.1007/s10806-016-9615-x

3 Napolitano, P., Pacelli, A., Girolami, A. and Braghieri, A. (2008) Effect of information about animal welfare on consumer willingness to pay for yogurt. *Journal of Dairy Science* 91: 910–17.

4 Harrison, R. (2013) *Animal Machines: The New Factory Farming Industry*, revised edition. CABI Publishing, Wallingford.

5 Agriculture and Horticultural Development Board (2021) UK and EU cow numbers. https://ahdb.org.uk/dairy/uk-and-eu-cow-numbers

6 Stolba, A. and Wood-Gush, D. (1989) The behaviour of pigs in a semi-natural environment. *Animal Science* 48(2): 419–25. https://doi.org/10.1017/S0003356100040411

7 Mutilations (Permitted Procedures) (England) Regulations 2007. Statutory Instrument no. 1100 of 2007. www.legislation.gov.uk/uksi/2007/1100/contents. Similar legislation applies in the other part of the United Kingdom.

8 Stafford, K.J. and Mellor, D.J. (2011) Addressing the pain associated with disbudding and dehorning in cattle. *Applied Animal Behaviour Science* 135: 226–31.

9 The geographical distribution of cattle movements appears to be relatively stable from year to year, with the great majority of animals moving less than 100 km per journey, although many tens of thousands move over far greater distances of up to 1,000 km. Mitchell, A., Bourn, D., Mawdsley, J. *et al.* (2005) Characteristics of cattle movements in Britain: an analysis of records from the Cattle Tracing System. *Animal Science* 80: 265–73.

10 UK Government (2020) Government consults on ending live animal exports for slaughter. Press release, 3 December 2020. www.gov.uk/government/news/government-consults-on-ending-live-animal-exports-for-slaughter

11 Agriculture and Horticulture Development Board (2020) Cattle and sheep abattoir size, and throughput risk in England. https://ahdb.org.uk/news/cattle-and-sheep-abattoir-size-and-throughput-risk-in-england

12 Raj, A.B.M. and Gregory, N.G. (1995) Welfare implications of gas stunning pigs 1. Determination of aversion to the initial inhalation of carbon dioxide or argon. *Animal Welfare* 4: 273–80.

13 European Food Safety Authority (2004) Opinion of the Scientific Panel on Animal Health and Welfare (AHAW) on a request from the Commission related to welfare aspects of the main systems of stunning and killing the main commercial species of animals. *EFSA Journal* 45: 1–29. https://doi.org/10.2903/j.efsa.2004.45

14 EFSA Panel on Animal Health and Welfare (2020) Scientific Opinion on the welfare of pigs at slaughter. *EFSA Journal* 18: 6148. https://doi.org/10.2903/j.efsa.2020.6148

15 Shields, S.J. and Raj, A.B. (2010) A critical review of electrical water-bath stun systems for poultry slaughter and recent developments in alternative technologies. *Journal of Applied Animal Welfare Science* 13: 281–99. https://doi.org/10.1080/10888705.2010.507119

16 British Veterinary Association (2019) Non-stun slaughter. www.bva.co.uk/take-action/our-policies/non-stun-slaughter

Chapter 4

1 St Kilda Soay Sheep Project (2022): https://soaysheep.bio.ed.ac.uk

2 Department for Environment, Food and Rural Affairs; Department of Agriculture, Environment and Rural Affairs (Northern Ireland); Welsh Government; Scottish Government (2020) *Agriculture in the United Kingdom 2019*. https://assets .publishing.service.gov.uk/government/uploads/system/uploads/attachment_data/ file/950618/AUK-2019-07jan21.pdf

3 OECD/FAO (2022) *OECD-FAO Agricultural Outlook (Edition 2021), OECD Agriculture Statistics (database)*. https://doi.org/10.1787/4bde2d83-en

4 Defra (2020) *Family Food 2018–2019*. www.gov.uk/government/statistics/family -food-201819/family-food-201819

5 OECD/FAO (2022): *OECD-FAO Agricultural Outlook (Edition 2021), OECD Agriculture Statistics (database)*

6 Hall, S.J.G. (1989) Chillingham cattle: social and maintenance behaviour in an ungulate that breeds all year round. *Animal Behaviour* 38: 215–25. https://doi.org /10.1016/s0003-3472(89)80084-3

7 Agriculture and Horticulture Development Board (2019) *The UK Cattle Yearbook 2019*. https://ahdb.org.uk/knowledge-library/the-uk-cattle-yearbook-2019

8 Defra (2003) *Code of Recommendations for the Welfare of Livestock: Cattle*. Defra, London. https://assets.publishing.service.gov.uk/government/uploads/system/ uploads/attachment_data/file/69368/pb7949-cattle-code-030407.pdf

9 Stafford, K.J. and Mellor, D.J. (2005) The welfare significance of the castration of cattle: a review. *New Zealand Veterinary Journal* 53: 271–78. https://doi.org/10 .1080/00480169.2005.36560

10 British Veterinary Association (2017) Analgesia in calves: policy statement. www.bva .co.uk/take-action/our-policies/analgesia-in-calf-disbudding-and-castration

11 March, M.D., Haskell, M.J., Chagunda, M.G.G., Langford, F.M. and Roberts, D.J. (2014) Current trends in British dairy management regimens. *Journal of Dairy Science* 97: 7985–94. https://doi.org/10.3168/jds.2014-8265

12 Agriculture and Horticultural Development Board (2021) UK and EU cow numbers. https://ahdb.org.uk/dairy/uk-and-eu-cow-numbers

13 House of Commons Library (2020) UK Dairy Industry Statistics. Briefing paper number 2721, 1 May 2020.

14 Bradley, A.J., Leach, K.A., Breen, J.E., Green, L.E. and Green, M.J. (2007) Survey of the incidence and aetiology of mastitis on dairy farms in England and Wales. *Veterinary Record* 160: 253–8. https://doi.org/10.1136/vr.160.8.253

15 Green, M.J., Leach, K.A., Breen, J.E., Green, L.E., and Bradley, A.J. (2007) National intervention study of mastitis control in dairy herds in England and Wales. *Veterinary Record* 160: 287–93. https://doi.org/10.1136/vr.160.9.287

16 Green *et al.* (2007) *Veterinary Record* 160: 287–93. https://doi.org/10.1136/vr.160 .9.287

17 Farm Animal Welfare Council (2009) Opinion on the welfare of the dairy cow. https://assets.publishing.service.gov.uk/government/uploads/system/ uploads/attachment_data/file/325044/FAWC_opinion_on_dairy_cow_welfare .pdf

18 Archer, A.C., Green, M.J. and Huxley, J.N. (2010) Association between milk yield and serial locomotion score assessments in UK dairy cows. *Journal of Dairy Science* 93: 4045–53. https://doi.org/10.3168/jds.2010-3062

19 Farm Animal Welfare Council (2009) Opinion on the welfare of the dairy cow.

20 Archer, S., Bell, N. and Huxley, J. (2010) Lameness in UK dairy cows: a review of the current status. *In Practice* 32: 492–504. https://doi.org/10.1136/inp.c6672

21 Clarkson, M.J., Downham, D.Y., Faull, W.B. *et al.* (1996) Incidence and prevalence of lameness in dairy cattle. *Veterinary Record* 138: 563–67. https://doi.org/10.1136/vr.138.23.563

22 Hare, E., Norman, H.D. and Wright, J.R. (2006) Survival rates and productive herd life of dairy cattle in the United States. *Journal of Dairy Science* 89: 3713–20. https://doi.org/10.3168/jds.S0022-0302(06)72412-2

23 Hughes, T. (1979) *Moortown Diary.* Faber and Faber, London.

24 Stafford, K.J. and Mellor, D.J. (2011) Addressing the pain associated with disbudding and dehorning in cattle. *Applied Animal Behaviour Science* 135: 226–31.

25 British Veterinary Association (2017) Analgesia in calves: policy statement.

26 Mueller, M.L., Cole, J.B., Connors, N.K. *et al.* (2021) Comparison of gene editing versus conventional breeding to introgress the POLLED allele into the tropically adapted Australian beef cattle population. *Frontiers in Genetics* 12: 68. https://doi.org/10.3389/fgene.2021.593154

27 Schuster, F., Aldag, P., Frenzel, A. *et al.* (2020) CRISPR/Cas12a mediated knock-in of the Polled Celtic variant to produce a polled genotype in dairy cattle. *Scientific Reports* 10: 13570. https://doi.org/10.1038/s41598-020-70531-y

28 Meagher, R.K., Beaver, A., Weary, D.M. and von Keyserlingk, M.A.G. (2019) Invited review: a systematic review of the effects of prolonged cow–calf contact on behavior, welfare, and productivity. *Journal of Dairy Science* 102: 5765–83. https://doi.org/10.3168/jds.2018-16021

29 Beaver, A., Meagher, R.K., von Keyserlingk, M.A.G. and Weary, D.M. (2019) Invited review: a systematic review of the effects of early separation on dairy cow and calf health. *Journal of Dairy Science* 102: 5784–810. https://doi.org/10.3168/jds.2018-15603

30 Flower, F.C. and Weary, D.M. (2003) Effects of early separation on the dairy cow and calf. *Animal Welfare* 12: 339–48. https://doi.org/10.1016/S0168-1591(00)00128-3

31 Levitt, T. (2019) Rise of ethical milk: 'mums ask when cows and their calves are separated'. *The Guardian*, 29 June 2019. www.theguardian.com/environment/2019/jun/29/mums-ask-when-cows-and-their-calves-separated-rise-ethical-milk-vegan

32 It should be noted that veal systems are now carefully regulated to prohibit the rearing of calves in crates and to prohibit the diets that induced anaemia in veal calves (which gives traditional veal its pale colour). Council of the European Union (2008) Council Directive 2008/119/EC of 18 December 2008. Laying down minimum standards for the protection of calves https://eur-lex.europa.eu/LexUriServ/LexUriServ.do?uri=OJ:L:2009:010:0007:0013:EN:PDF.

33 Levitt, T. (2018) Dairy's 'dirty secret': it's still cheaper to kill male calves than to rear them. *The Guardian*, 26 March 2018: www.theguardian.com/environment/2018/mar/26/dairy-dirty-secret-its-still-cheaper-to-kill-male-calves-than-to-rear-them

34 Agriculture and Horticulture Development Board (2019) Reduce costs and increase profits by using sexed semen. https://ahdb.org.uk/news/reduce-costs-and-increase-your-potential-profit-by-using-sexed-semen

35 Compassion in World Farming (undated) Dairy cows. www.ciwf.org.uk/farm
    -animals/cows/dairy-cows

36 Farm Animal Welfare Council (2007) Report on stockmanship and farm animal
    welfare. https://assets.publishing.service.gov.uk/government/uploads/system/
    uploads/attachment_data/file/325176/FAWC_report_on_stockmanship_and_farm
    _animal_welfare.pdf.

37 Gieseke, D., Lambertz, C. and Gauly, M. (2018) Relationship between herd size
    and measures of animal welfare on dairy cattle farms with freestall housing in
    Germany. *Journal of Dairy Science* 101: 7397–411. https://doi.org/10.3168/jds
    .2017-14232

38 Arnott, G., Ferris, C. and O'Connell, N. (2017) Review: welfare of dairy cows in
    continuously housed and pasture-based production systems. *Animal* 11: 261–273.
    https://doi.org/10.1017/S1751731116001336

39 Keyserlingk, M., Olenick, D. and Weary, D. (2008) Acute behavioral effects of
    regrouping dairy cows. *Journal of Dairy Science* 91: 1011–16. https://doi.org/10
    .3168/jds.2007-0532

40 Defra (2021) Livestock numbers in England and the UK: annual statistics on the
    number of livestock in England and the UK in June and December. www.gov.uk/
    government/statistical-data-sets/structure-of-the-livestock-industry-in-england-at
    -december

41 Mutilations (Permitted Procedures) (England) Regulations 2007. Statutory
    Instrument no. 1100 of 2007. www.legislation.gov.uk/uksi/2007/1100/contents

42 Gascoigne, E., Mouland, C. and Lovatt, F. (2021) Considering the 3Rs for
    castration and tail docking in sheep. *In Practice* 43: 152–62. https://doi.org/10
    .1002/inpr.29

43 British Veterinary Association (2020) Sheep castration, tail docking, and pain
    management: policy statement. www.bva.co.uk/media/3364/sheep-castration-tail
    -docking-and-pain-management-final.pdf

44 Mutilations (Permitted Procedures) (England) Regulations 2007.

45 Gascoigne, E., Mouland, C. and Lovatt, F. (2021) *In Practice* 43: 152–62.

46 British Veterinary Association (2020) Sheep castration, tail docking, and pain
    management: policy statement.

47 Winter, J.R. and Green, L.E. (2017) Cost–benefit analysis of management practices
    for ewes lame with footrot. *Veterinary Journal* 220: 1–6. https://doi.org/10.1016/j
    .tvjl.2016.11.010

48 Farm Animal Welfare Council (2011) Opinion on lameness in sheep. https://assets
    .publishing.service.gov.uk/government/uploads/system/uploads/attachment_data/
    file/325039/FAWC_opinion_on_sheep_lameness.pdf

49 Farm Animal Welfare Council (2011) Opinion on lameness in sheep.

50 Winter, J.R., Kaler, J., Ferguson, E., KilBride, A.L. and Green, L.E. (2015) Changes
    in prevalence of, and risk factors for, lameness in random samples of English sheep
    flocks: 2004–2013. *Preventive Veterinary Medicine* 122: 121–8. https://doi.org/10
    .1016/j.prevetmed.2015.09.014

51 Mellor, D.J., Beausoleil, N.J., Littlewood, K.E. *et al.* (2020) The 2020 five domains
    model: including human–animal interactions in assessments of animal welfare.
    Animals 10: 1870. https://doi.org/10.3390/ani10101870

Chapter 5

1 Reardon, S. (2022) First pig-to-human heart transplant: what can scientists learn? *Nature* 601: 305–6. https://doi.org/10.1038/d41586-022-00111-9

2 Winston Churchill Quotes. www.goodreads.com/quotes/214859-always-remember-a-cat-looks-down-on-man-a-dog

3 Croney, C.C. and Boysen, S.T. (2021) Acquisition of a joystick-operated video task by pigs (*Sus scrofa*). *Frontiers in Psychology* 12: 631755. https://doi.org/10.3389/fpsyg.2021.631755

4 OECD/FAO (2022) OECD-FAO Agricultural Outlook (Edition 2021), OECD Agriculture Statistics (database). https://doi.org/10.1787/4bde2d83-en

5 OECD/FAO (2022) OECD-FAO Agricultural Outlook (Edition 2021), OECD Agriculture Statistics (database).

6 Gethöffer, F., Sodeikat, G. and Pohlmeyer, K. (2007) Reproductive parameters of wild boar (*Sus scrofa*) in three different parts of Germany. *European Journal of Wildlife Research* 53: 287–97. https://doi.org/10.1007/s10344-007-0097-z

7 Schedule 3 of the Mutilations (Permitted Procedures) (England) Regulations 2007. Statutory Instrument no. 1100 of 2007. www.legislation.gov.uk/uksi/2007/1100/contents

8 Sinclair, A.R., Tallet, C., Renouard, A. *et al.* (2019) Behaviour of isolated piglets before and after tooth clipping, grinding or sham-grinding. Abstract from 53rd Congress of the International Society for Applied Ethology (ISAE), Bergen, Norway.

9 Hay, M., Rue, J., Sansac, C., Brunel, G. and Prunier A. (2004) Long-term detrimental effects of tooth clipping or grinding in piglets: a histological approach. *Animal Welfare* 13: 27–32.

10 Defra (2020) *Code of Practice for the Welfare of Pigs*. https://assets.publishing.service.gov.uk/government/uploads/system/uploads/attachment_data/file/908108/code-practice-welfare-pigs.pdf

11 Welfare of Livestock Regulations 1994. Statutory Instrument no. 2126 of 1994. www.legislation.gov.uk/uksi/1994/2126/introduction/made. Mutilations (Permitted Procedures) (England) Regulations 2007. Statutory Instrument no. 1100 of 2007. www.legislation.gov.uk/uksi/2007/1100/contents

12 Defra (2020) *Code of Practice for the Welfare of Pigs*.

13 Tallet, C., Rakotomahandry, M., Herlemont, S. and Prunier, A. (2019) Evidence of pain, stress, and fear of humans during tail docking and the next four weeks in piglets (*Sus scrofa domesticus*). *Frontiers in Veterinary Science* 6: 462. https://doi.org/10.3389/fvets.2019.00462

14 Pandolfi, F., Stoddart, K., Wainwright, N., Kyriazakis, I. and Edwards, S. (2017) The 'Real Welfare' scheme: benchmarking welfare outcomes for commercially farmed pigs. *Animal* 11: 1816–24.

15 De Briyne, N., Berg, C., Blaha, T., Palzer, A. and Temple, D. (2018) Phasing out pig tail docking in the EU – present state, challenges and possibilities. *Porcine Health Management* 4: 27. https://doi.org/10.1186/s40813-018-0103-8

16 Horrell, R.I., A'Ness, P.J., Edwards, S. and Eddison, J. (2001) The use of nose-rings in pigs: consequences for rooting, other functional activities, and welfare. *Animal Welfare* 10: 3–22.

17 Defra (2020) *Code of Practice for the Welfare of Pigs*.

18 Mellor, D.J., Beausoleil, N.J., Littlewood, K.E. *et al.* (2020) The 2020 Five Domains Model: including human–animal interactions in assessments of animal welfare. *Animals* 10: 1870. https://doi.org/10.3390/ani10101870

19 Webster, J. (2005) *Animal Welfare: Limping Towards Eden.* Blackwell, Oxford.

20 The Hisex Brown (https://www.hisex.com/en/product/hisex-brown) is one of the most popular and successful hybrid laying hens in the world. The Ross 308 hybrid (http://eu.aviagen.com/brands/ross) is a typical modern fast-growing broiler chicken.

21 Collias, N.E. and Collias, E.C. (1967) A field study of the Red Jungle Fowl in north-central India. *The Condor* 69: 360–86.

22 Menzies, F.D., Goodall, E.A., McConachy, A. and Alcorn M.J. (1998) An update on the epidemiology of contact dermatitis in commercial broilers. *Avian Pathology* 27: 174–80.

23 Knowles, T.G., Kestin, S.C., Haslam, S.M. *et al.* (2008) Leg disorders in broiler chickens: prevalence, risk factors and prevention. *PLOS One* 3(2): e1545. https://doi.org/10.1371/journal.pone.0001545

24 Commission of the European Communities, Health and Consumer Protection Directorate General (2000) *The Welfare of Chickens Kept for Meat Production (Broilers).* Report of the Scientific Committee on Animal Health and Animal Welfare, Commission of the European Communities, Brussels. https://ec.europa.eu /food/system/files/2020-12/sci-com_scah_out39_en.pdf

25 Dawkins, M.S., Donnelly, C.A. and Jones, T.A. (2004) Chicken welfare is influenced more by housing conditions than by stocking density. *Nature* 427: 342–4.

26 European Union (2007) Council Directive 2007/43/EC of 28 June 2007 laying down minimum rules for the protection of chickens kept for meat production. https://eur-lex.europa.eu/legal-content/EN/TXT/?uri=celex%3A32007L0043

27 In the UK, when stocking above 33 kg per square metre, documents with technical details and information on equipment must be kept in the housing, including: a plan that shows the dimensions of the surfaces the chickens occupy; a ventilation plan and target air quality levels (including airflow, air speed and temperature) and details of cooling and heating systems and their location; the location and nature of feeding and watering systems (e.g. automatic or manual, how many feeders, how each is operated); alarm and backup systems if any equipment essential for the chickens' health and wellbeing fails; floor type and litter normally used; records of technical inspections of the ventilation and alarm systems. Defra (2019) Broiler (meat) chickens: welfare recommendations. www .gov.uk/government/publications/poultry-on-farm-welfare/broiler-meat-chickens -welfare-recommendations

28 The Better Chicken Commitment. https://betterchickencommitment.com/en-GB/ policy

29 De Jong, I.C. and Guémené, D. (2011) Major welfare issues in broiler breeders. *World's Poultry Science Journal* 67: 73–82. https://doi.org/10.1017/ S0043933911000067

30 Royal Society for the Prevention of Cruelty to Animals (2021) Laying hens – farming (egg production). www.rspca.org.uk/adviceandwelfare/farm/layinghens/farming

31 Defra (2021) Quarterly UK statistics about eggs – statistics notice (data to June 2021). www.gov.uk/government/statistics/egg-statistics/quarterly-uk-statistics-about -eggs-statistics-notice-data-to-june-2021

32 Shoppers urged to buy white eggs because they come from less-aggressive hens to end cruel practice of beak trimming. *Daily Mail*, 7 March 2020. www.dailymail.co .uk/news/article-8086869/Shoppers-urged-buy-white-eggs-come-aggressive-hens .html

33 In the UK, 'free range' represented 63% of egg production in 2020 (www.gov.uk /government/statistics/egg-statistics/quarterly-uk-statistics-about-eggs-statistics -notice-data-to-june-2021). In the EU, free range represents less than 12% (2019 figures, https://ec.europa.eu/info/food-farming-fisheries/animals-and-animal -products/animal-products/eggs_en).

34 De Boer, I.J.M. and Cornelissen, A.M.G. (2002) A method using sustainability indicators to compare conventional and animal-friendly egg production systems. *Poultry Science* 81: 173–81. https://doi.org/10.1093/ps/81.2.173

35 In 1990 it was reported that 75% of all commercial layers in the world and 95% in the United States were kept in cages. North, M.O. and Bell, D.E. (1990) *Commercial Chicken Production Manual*, 4th edition, pp. 297, 315. Van Nostrand Reinhold, New York.

36 Harrison, R. (2013) *Animal Machines: The New Factory Farming Industry*, revised edition. CABI Publishing, Wallingford. First edition published in 1964.

37 Commission of the European Communities, Health and Consumer Protection Directorate General (1996) *Report of the Scientific Veterinary Committee Animal Welfare Section on the Welfare of Laying Hens*. Commission of the European Communities, Brussels.

38 Cooper, J.J. and Appleby, M.C. (1993) Quantifying nesting motivation in domestic hens. *Proceedings of the British Society of Animal Production (1972)* 1993: Winter meeting, March 1993, p. 78. https://doi.org/10.1017/ S0308229600024041

39 Compassion in World Farming (2021) Welfare issues for egg laying hens. www.ciwf .org.uk/farm-animals/chickens/egg-laying-hens

40 Savory, J. (2004) Laying hen welfare standards: a classic case of 'power to the people'. *Animal Welfare* 13: 153–8.

41 Fleming, R.H., McCormack, H.A., McTeir, L. and Whitehead, C.C. (2006) Relationships between genetic, environmental and nutritional factors influencing osteoporosis in laying hens. *British Poultry Science* 47: 742–55. https://doi.org/10 .1080/00071660601077949

42 Farm Animal Welfare Council (2010) Opinion on Osteoporosis and Bone Fractures in Laying Hens. www.gov.uk/government/publications/fawc-opinion-on-osteopo- rosis-and-bone-fractures-in-laying-hens

43 Gregory, N.G. and Wilkins, L.J. (1989) Broken bones in domestic fowl: handling and processing damage in end-of-lay battery hens. *British Poultry Science* 30: 555–62. https://doi.org/10.1080/00071668908417179

44 Webster, J.B. (2004) Welfare implications of avian osteoporosis. *Poultry Science* 83: 184–92. https://doi.org/10.1093/ps/83.2.184

45 Birkhead, T. (2012) *Bird Sense: What it's Like to be a Bird*. Bloomsbury, London.

46 Defra (2018) *Code of Practice for the Welfare of Laying Hens and Pullets*, paragraph 100. HMSO, London. https://assets.publishing.service.gov.uk/government/uploads /system/uploads/attachment_data/file/732227/code-of-practice-welfare-of-laying -hens-pullets.pdf

47 Porter, R. (2019) Analysis: Increasing pressure to ban infra-red beak treatment. *Poultry News* 13, February 2019. www.poultrynews.co.uk/health-welfare/welfare/analysis-increasing-pressure-to-ban-infra-red-beak-treatment.html.

48 Guhl, A.M. (1953) The social behaviour of the domestic fowl. *Technical Bulletin of the Agricultural Experimental Station, Kansas State College* No. 73.

49 Hughes, B.O. (1975) The concept of an optimum stocking density and its selection for egg production. In *Economic Factors Affecting Egg Production* (ed. B.M. Freeman and K.N. Boorman). *Poultry Science Symposium* 10: 271–98.

50 Appleby, M.C. and Hughes, B.O. (1991) Welfare of laying hens in cages and alternative systems: environmental, physical and behavioural aspects. *World's Poultry Science Journal* 47: 109–28. https://doi.org/10.1079/wps19910013

51 Council of the European Union (1999) Council Directive 1999/74/EC of 19 July 1999.

52 Soil Association (undated) Organic vs. free-range – what's the difference? www.soilassociation.org/organic-living/what-is-organic/organic-eggs

53 FAOSTATS (2019) Production: Livestock primary, Rome. Food and Agriculture Organization of the United Nations, Statistics Division.

54 RSPCA (2015) Watertight: the case for providing farmed ducks with full body access to water. www.rspca.org.uk/adviceandwelfare/farm/ducks

55 Jones, T.A., Waitt, C. and Dawkins, M.S. (2009) Water off a duck's back: showers and troughs match ponds for improving duck welfare. *Applied Animal Behaviour Science* 116: 52–7.

56 Ruis, M. and van Krimpen, M. (2011) Open water provision for Pekin ducks to increase natural behaviour requires an integrated approach. *Proceedings of the 45th Congress of the International Society for Applied Ethology, Indianapolis, 31 July–4 August 2011* (ed. E.A. Pajor and J.N. Marchant-Forde). Wageningen Academic Publishers, Wageningen.

57 Scottish Government (2020) Scottish Fish Farm Production Survey 2019. www.gov.scot/publications/scottish-fish-farm-production-survey-2019

58 Statista (2020) Global fish production from 2002 to 2019. www.statista.com/statistics/264577/total-world-fish-production-since-2002

59 Farm Animal Welfare Committee (2014) Opinion on the welfare of farmed fish. FAWC, London. https://assets.publishing.service.gov.uk/government/uploads/system/uploads/attachment_data/file/319323/Opinion_on_the_welfare_of_farmed_fish.pdf

60 Compassion in World Farming (undated) Fish welfare. www.ciwf.org.uk/farm-animals/fish/fish-welfare

61 Ashley, P.J. (2007) Fish welfare: current issues in aquaculture. *Applied Animal Behaviour Science* 104: 199–235. https://doi.org/10.1016/j.applanim.2006.09.001.

62 Samsing, F., Oppedal, F., Johansson, D., Bui, S. and Dempster, T. (2014) High host densities dilute sea lice *Lepeophtheirus salmonis* loads on individual Atlantic salmon, but do not reduce lice infection success. *Aquaculture Environment Interactions* 6: 81–9. https://doi.org/10.3354/aei00118

Chapter 6

1 Rackham, O. (1986) *The History of the Countryside: The Full Fascinating Story of Britain's Landscape.* J.M. Dent & Sons, London.

2 Two excellent examples of books about wildlife loss and how it might be restored are: (a) Tree, I. (2018) *Wilding: The Return of Nature to a British Farm.* Picador, London; (b) Macdonald, B. (2019) *Rebirding: Rewilding Britain and its Birds,* Pelagic Publishing, Exeter.

3 (a) Animal Welfare Act (2006) www.legislation.gov.uk/ukpga/2006/45/contents. The act applies to England and Wales only. Similar but not identical acts apply in Scotland and Northern Ireland. (b) Natural England (2019) Wildlife management advice note: the Animal Welfare Act 2006. https://assets.publishing.service.gov.uk/government/uploads/system/uploads/attachment_data/file/901837/wml-gu02-animal-welfare-act-wildlife-managment.pdf

4 Wild Mammals (Protection) Act 1996 www.legislation.gov.uk/ukpga/1996/3/contents.

5 Wildlife and Countryside Act 1981 www.legislation.gov.uk/ukpga/1981/69.

6 Protection of Badgers Act 1992 www.legislation.gov.uk/ukpga/1992/51/contents.

7 Natural England (2021) General licences for wildlife management. www.gov.uk/government/collections/general-licences-for-wildlife-management

8 Natural England (2021) Wildlife licences: when you need to apply. www.gov.uk/guidance/wildlife-licences

9 Doward, J. (2020) More than 100,000 badgers slaughtered in discredited cull policy. *The Guardian*, 28 March 2020. www.theguardian.com/environment/2020/mar/28/more-than-100000-badgers-slaughtered-in-discredited-cull-policy

10 Natural England (2021) Summary of 2020 badger control operations www.gov.uk/government/publications/bovine-tb-summary-of-badger-control-monitoring-during-2020/summary-of-2020-badger-control-operations.

11 Independent Expert Panel (2014) *Pilot Badger Culls in Somerset and Gloucestershire* Report by the Independent Expert Panel, Chair: Professor Ranald Munro. Presented to the Secretary of State for Environment, Food and Rural Affairs, The Rt Hon Owen Paterson MP, 20 March. http://assets.publishing.service.gov.uk/government/uploads/system/uploads/attachment_data/file/300382/independent-expert-panel-report.pdf

12 Data from a series of Natural England publications summarising badger control operations between 2013 and 2020, the latest of which is: Natural England (2021) Summary of 2020 badger control operations.

13 Natural England (2021) Summary of 2020 badger control operations and earlier reports in the series.

14 British Veterinary Association (2015) BVA calls for change to badger culling method and wider roll-out in England. www.bva.co.uk/news-and-blog/news-article/bva-calls-for-change-to-badger-culling-method-and-wider-roll-out-in-england

15 Natural England (2021) Summary of 2020 badger control operations.

16 Information Commissioner (2017) Freedom of Information Act 2000 (FOIA) Environmental Information Regulations 2OO4 (EIR) Decision notice reference: FER0659789. https://ico.org.uk/media/action-weve-taken/decision-notices/2017/2014420/fer0659789.pdf

17 Approved traps are listed on separate Orders across the UK administrations:

• Spring Traps Approval (England) Order 2018, made under the Pests Act 1954
• Spring Traps Approval (Wales) Order 2019, made under the Pests Act 1954

- Spring Traps Approval (Scotland) Order 2011 (as amended by the Spring Traps Approval (Scotland) Amendment Order 2018), made under the Agriculture (Scotland) Act 1948
- Spring Traps Approval Order (Northern Ireland) 2019, made under the Wildlife (Northern Ireland) Order 1985

The UK is a signatory to the Agreement on International Humane Trapping Standards (AIHTS – https://eur-lex.europa.eu/collection/eu-law/treaties/treaties-force .html). More recently, the Humane Trapping Standards Regulations 2019 and the Humane Trapping Standards Regulations (Northern Ireland) 2019 were introduced to implement the requirements of the AIHTS for relevant species found in the UK (stoats), permitting the use of certain specified types of trap under general licence.

18 Harris, S. and Thain, B. (2020) *Hanged by the Feet Until Dead: an Analysis of Snaring and Trapping on Scottish Grouse Moors*. A report commissioned by the Director of the League Against Cruel Sports Scotland. https://raptorpersecutionsc otland.files.wordpress.com/2020/08/hanged-by-the-feet-until-dead-3.pdf

19 A review of the use of live traps for birds produced by the then Scottish Natural Heritage (Now NatureScot) includes detailed descriptions of each type of trap: Campbell, S.T., Hartley, F.G. and Reynolds, J.C. (2016) Assessing the nature and use of corvid cage traps in Scotland: Part 4 of 4 – Review and recommendations. Scottish Natural Heritage Commissioned Report No. 934.

20 NatureScot (2021) Trap registration. www.nature.scot/professional-advice/protected -areas-and-species/licensing/trap-registration

21 Baker, S.E., Sharp, T.M. and Macdonald, D.W. (2016) Assessing animal welfare impacts in the management of European rabbits (*Oryctolagus cuniculus*), European moles (*Talpa europaea*) and Carrion crows (*Corvus corone*). *PLOS One* 11(1): e0146298. https://doi.org/10.1371/journal.pone.0146298

22 In the absence of data about the stress of enforced confinement and the handling of corvids, a comparison with other species of birds is about the best we can do. Geese when handled only for five minutes showed a dramatic increase in the level of humoral indices of stress (that is, levels of hormones released into the bloodstream during stressful experiences). These increased several-fold within two minutes, and the return to initial values could take up to 1 hour. Le Maho, Y., Karmann, H., Briot, H. *et al.* (1992) Stress in birds due to routine handling and a technique to avoid it. *American Journal of Physiology. Regulatory, Integrative and Comparative Physiology* 32: 775–81. https://doi.org/10.1152/ajpregu.1992.263.4.r775

23 Report of the Independent Working Group on Snares (2005). https://webarchive .nationalarchives.gov.uk/ukgwa/20130402151656/http://archive.defra.gov.uk/ wildlife-pets/wildlife/management/documents/snares-iwgs-report.pdf

24 Section 13 of the Wildlife and Natural Environment (Scotland) Act 2011. www .legislation.gov.uk/asp/2011/6/contents

25 Section 11A of the Wildlife and Countryside Act 1981, as amended by Section 13 of the Wildlife and Natural Environment (Scotland) Act 2011. www .legislation.gov.uk/ukpga/1981/69/section/11F

26 Report of the Independent Working Group on Snares (2005).

27 OneKind (2010) *The OneKind Report on Snaring*. www.snarewatch.org/images/ resources/onekind_report_snaring_2010.pdf

28 Report of the Independent Working Group on Snares (2005).
29 Report of the Independent Working Group on Snares (2005).
30 Defra (2005) *Defra Code of Practice on the use of snares in fox and rabbit control.* http://adlib.everysite.co.uk/resources/000/125/893/snares-cop.pdf
31 Defra (2009) Determining the extent of use and humaneness of snares in England and Wales. Defra Science Report WM 0315. http://sciencesearch.defra.gov.uk /Default.aspx?Menu=Menu&Module=More&Location=None&Completed=0 &ProjectID=14689.
32 Harris, S. (2022) A review of the use of snares in the UK. A report prepared for Animal Aid, April 2022. https://www.animalaid.org.uk/wp-content/ uploads/2022/04/Snaring-report-final-version.pdf.
33 Guild of British Molecatchers (undated) Traditional molecatchers code of working practice. www.guildofbritishmolecatchers.co.uk/?p=code.of.practice
34 (a) Baker, S.E. and Macdonald D.W. (2012) Not so humane mole tube traps. *Animal Welfare* 21: 613–15. (b) Baker, S.E., Shaw, R.F., Atkinson, R.P.D., West, P. and Macdonald, D.W. (2015) Potential welfare impacts of kill-trapping European moles (*Talpa europaea*) using scissor traps and Duffus traps: a post-mortem examination study. *Animal Welfare* 24: 1–14. https://doi.org/10.7120/09627286.24.1.001
35 Atkinson, R.P.D., Macdonald, D. and Johnson, P. (2008) The status of the European Mole *Talpa europaea* L. as an agricultural pest and its management. *Mammal Review* 24: 73–90. https://doi.org/10.1111/j.1365-2907.1994.tb00136.x
36 Mason, G.J. and Littin, K.E. (2003) The humaneness of rodent pest control. *Animal Welfare* 12: 1–37.
37 Anglers urging government to allow shooting of more cormorants to stop birds eating fish they catch for sport. *The Independent*, 18 August 2018. www .independent.co.uk/climate-change/news/angligs-trust-fishing-cormorants-shooting -licence-rspb-a8489961.html
38 Östman, Ö., Bergenius, M., Boström, M.K. and Lunneryd, S.-G. (2012) Do cormorant colonies affect local fish communities in the Baltic Sea? *Canadian Journal of Fisheries and Aquatic Sciences* 69: 1047–55. https://doi.org/10.1139/f2012-042
39 Engström, H. (2001) Long term effects of cormorant predation on fish communities and fishery in a freshwater lake. *Ecography* 24: 127–38.
40 Ovegård, M.K., Jepsen, N., Bergenius, M. and Petersson, E. (2021) Cormorant predation effects on fish populations: a global meta-analysis. *Fish and Fisheries* 22: 605–22. https://doi.org/10.1111/faf.12540
41 Natural England (2020) Summary of wildlife licences issued by Natural England in 2019. www.gov.uk/government/publications/summary-of-wildlife-licences-issued -by-natural-england-in-2019
42 Defra (2013) Evidence summary: Review of fish-eating birds policy, 19 July 2013. https://assets.publishing.service.gov.uk/government/uploads/system/uploads/ attachment_data/file/224186/pb13972-fish-eating-birds-evidence-130719.pdf
43 (a) Baker, S.E., Ellwood, S.A., Tagarielli, V.L. and Macdonald, D.W. (2012) Mechanical performance of rat, mouse and mole spring traps, and possible implications for welfare performance. *PLOS One* 7(6): e39334. (b) Baker, S.E., Macdonald, D.W. and Ellwood, S.A. (2017) Double standards in spring trap welfare: ending inequality for rats (Rodentia: Muridae), mice (Rodentia: Muridae) and moles (Insectivora: Talpidae) in the United Kingdom. In *Proceedings of the*

Ninth International Conference on Urban Pests (ed. M.P. Davies, C. Pfeiffer and W.H. Robinson). Pureprint Group, Sussex, pp. 139–45.

44 Pest Management Alliance (undated) Humane use of rodent glue boards: code of practice. www.pmalliance.org.uk/codes-of-best-practice

45 Ministry of Primary Industries (2015) Glueboard traps prohibited from 2015. MPI Media Release www.mpi.govt.nz/news/media-releases/glueboard-traps-prohibited-from-2015/

46 (a) 'Crude and horrific' rodent glue traps may be banned. *The Times*, 27 October 2017. www.thetimes.co.uk/article/crude-and-horrific-rodent-glue-traps-may-be-banned-sxjcsb68s. (b) Glue Traps (Offences) Act 2022. https://www.legislation.gov.uk/ukpga/2022/26/enacted

47 British Veterinary Association (2019) Rodent glue traps: policy statement. www.bva.co.uk/media/1159/rodent-glue-traps.pdf

48 Mason, G.M. and Littin, K.E. (2003) The humaneness of rodent pest control. *Animal Welfare* 12: 1–37.

49 Ministry of Agriculture, Fisheries and Food (1997) Assessment of humaneness of vertebrate control agents. Evaluation of fully approved or provisionally approved products under the Control of Pesticides Regulations 1986. Pesticide Safety Directorate, MAFF, London. Available at www.pesticides.gov.uk/psd_evaluation_all.asp.

50 Baker, S,E., Ayers M., Beausoleil, N.J. *et al.* (2022) An assessment of animal welfare impacts in wild Norway rat (*Rattus norvegicus*) management. *Animal Welfare* 31: 51–68. https://doi.org/10.7120/09627286.31.1.005

51 Universities Federation for Animal Welfare (undated) Humane rodent and mole control: guiding principles in the humane control of rats and mice. www.ufaw.org.uk/rodent-welfare/rodent-welfare

52 Clayton, R. and Cowan, P. (2010) Management of animal and plant pests in New Zealand: patterns of control and monitoring by regional agencies. *Wildlife Research* 37: 360–71.

Chapter 7

1 From the homepage of The Wildlife Trusts. https://www.wildlifetrusts.org/about-us

2 Wallach, A.D., Bekoff, M., Batavia, C., Nelson, M.P. and Ramp, D (2018) Summoning compassion to address the challenges of conservation. *Conservation Biology* 32: 1255–65. https://doi.org/10.1111/cobi.13126

3 Oommen, M.A., Cooney, R., Ramesh, M. *et al.* (2019) The fatal flaws of compassionate conservation. *Conservation Biology* 33: 784–7. https://doi.org/10.1111/cobi.13329

4 Frey, S.N., Conover, M.R. and Cook, G. (2007) Successful use of neck snares to live-capture red foxes, *Human–Wildlife Interactions* 1.1, Article 10. https://doi.org/10.26077/r2wd-c109

5 Harper, M. (2018) The conservationist's dilemma: an update on the science, policy and practice of the impact of predators on wild birds (5). https://community.rspb.org.uk/ourwork/b/martinharper/posts/the-conservationist-39-s-dilemma-an-update-on-the-science-policy-and-practice-of-the-impact-of-predators-on-wild-birds-5

6 McKenzie, R. (2014) *Islay Sustainable Goose Management Strategy, October 2014 – April 2024*. Nature Scotland. www.nature.scot/sites/default/files

/2017-07/A1434517%20-%20ISLAY%20SUSTAINABLE%20GOOSE
%20MANAGEMENT%20STRATEGY%202014%20-%202024%20-
%20October%202014%20%28A2332648%29.pdf

7 RSPB and WWT (undated) Complaint to the Commission of the European
Communities concerning failure to comply with Community law. http://ww2.rspb
.org.uk/Images/goosecomplaint_tcm9-407227.pdf

8 Percival, S. and Bignal, E. (2018) The Islay Barnacle Goose management strategy: a
suggested way forward. *British Wildlife* 30: 37–44.

9 Duarte, L.M.G. (2013) Impacts of capture and handling on wild birds. PhD thesis,
Cardiff University.

10 Clewley, G.D., Robinson, R.A. and Clark, J.A. (2018) Estimating mortality rates
among passerines caught for ringing with mist nets using data from previously
ringed birds. *Ecology and Evolution* 8: 5164–72. https://doi.org/10.1002/ece3.4032.

11 Griesser, M., Schneider, N.A., Collis, M-A. *et al.* (2012) Causes of ring-related leg
injuries in birds: evidence and recommendations from four field studies. *PLoS One*
7(12): e51891. https://doi.org/10.1371/journal.pone.0051891

12 Geen, G.R., Robinson, R.A. and Baillie, S.R. (2019) Effects of tracking devices on
individual birds: a review of the evidence. *Journal of Avian Biology*. https://doi.org
/10.1111/jav.01823

13 Peniche, G., Vaughan-Higgins, R., Carter, I. *et al.* (2011) Long-term health effects
of harness-mounted radio transmitters in red kites (*Milvus milvus)* in England.
Veterinary Record 169: 311. https://doi.org/10.1136/vr.d4600

14 Thaxter, C.B., Ross-Smith, V.H., Clark, J.A. *et al.* (2016) Contrasting effects of GPS
device and harness attachment on adult survival of Lesser Black-backed Gulls *Larus fuscus*
and Great Skuas *Stercorarius skua*. *Ibis* 158: 279–90. https://doi.org/10.1111/ibi.12340

15 Sergio, F., Tavecchia, G., Tanferna, A. *et al.* (2015) No effect of satellite tagging
on survival, recruitment, longevity, productivity and social dominance of a raptor,
and the provisioning and condition of its offspring. *Journal of Applied Ecology* 52:
1665–75. https://doi.org/10.1111/1365-2664.12520

16 Curk, T., Scacco, M., Safi, K. and others. (2021) Wing tags severely impair
movement in African Cape Vultures. *Animal Biotelemetry* 9: 11. https://doi.org/10
.1186/s40317-021-00234-2

17 Moorhouse, T. and Macdonald, D. (2005). Indirect negative impacts of radio-
collaring: sex ratio variation in water voles. *Journal of Applied Ecology* 42: 91–98.
Doi: 10.1111/j.1365-2664.2005.00998.x

18 (a) Baker, G., Lumsden, L., Dettmann, E. *at al.* (2001) The effect of forearm bands
on insectivorous bats (Microchiroptera) in Australia. *Wildlife Research* 28. 229–37.
https://doi.org/10.1071/WR99068. (b) Ellison, L.E. (2008) *Summary and Analysis
of the U.S. Government Bat Banding Program*. US Geological Survey Open-File
Report 2008-1363. https://pubs.usgs.gov/of/2008/1363/pdf/OF08-1363_508.pdf

19 Collins J (2016) *Bat Surveys for Professional Ecologists: Good Practice Guidelines*, 3rd
edition. Bat Conservation Trust, London.

20 Sanchez-Ortiz, K., Gonzalez, R. E., De Palma, A. *et al.* (2019) Land-use and related
pressures have reduced biotic integrity more on islands than on mainlands. https://
doi.org/10.1101/576546

21 Goddard, P. (undated) Animal reintroductions: who is safeguarding animal
welfare? Wild Animal Welfare Committee, Topic paper no. 1. https://static1

.squarespace.com/static/5edf4fd72d25275e3acc8c4a/t/5f4f99b13c28b129b0df37f3
/1599052212900/Topic_Statement_Reintroductions_Final_Feb_2020.pdf

22 International Union for the Conservation of Nature (2013) *Guidelines for Reintroductions and Other Conservation Translocations*, Version 1.0. www.iucn.org/content/guidelines-reintroductions-and-other-conservation-translocations

23 Jeffs, C., Davies, M., Carter, I. *et al.* (2016) Reintroducing the Cirl Bunting to Cornwall. *British Birds* 109: 374–88.

24 Pullar, P. (2019) The buzzard is a success story – but it is too successful? *Scottish Field*, 6 September 2019. www.scottishfield.co.uk/outdoors/wildlifeandconservation/the-buzzard-is-a-success-story-but-it-is-too-successful

25 (a) Natural England (2019) Latest update: Wild bird licensing. https://naturalengland.blog.gov.uk/2019/04/08/latest-update-wild-bird-licensing-2. (b) Natural England (2016) Licence: Kill or take wild birds or their eggs, use a prohibited method or disturb wild birds or their nests when in use or being built. https://assets.publishing.service.gov.uk/government/uploads/system/uploads/attachment_data/file/551012/5._2016-_24189-SPM-WLM__Licence_Redacted.pdf

26 SongBird Survival website. www.songbird-survival.org.uk

27 Balmer, D., Gillings, S., Caffrey, B. *et al.* (2013) *Bird Atlas 2007–11: The Breeding and Wintering Birds of Britain and Ireland*. BTO, Thetford.

28 Newton, I., Dale, L. and Rothery, P. (1997) Apparent lack of impact of Sparrowhawks on the breeding densities of some woodland songbirds. *Bird Study* 44: 129–35. https://doi.org/10.1080/00063659709461048

29 Perrins, C.M. and Geer, T.A. (1980) The effect of sparrowhawks on tit populations. *Ardea* 68: 133–42.

30 Macdonald, M.A. and Bolton, M. (2008) Predation on wader nests in Europe. *Ibis* 150 (Suppl. 1): 54–73.

31 Tapper, S.C., Potts, G.R. and Brockless, M.H. (1996) The effect of an experimental reduction in predation pressure on the breeding success and population density of grey partridge *Perdix perdix*. *Journal of Applied Ecology* 33: 965–78.

32 Woodward, I.D., Massimino, D., Hammond, M.J. *et al.* (2020) BirdTrends 2020: trends in numbers, breeding success and survival for UK breeding birds. BTO Research Report 732. BTO, Thetford. www.bto.org/birdtrends

33 Zielonka, N.B., Hawkes, R.B., Jones, H., Burnside, R.J. and Dolman, P.M. (2019) Placement, survival and predator identity of Eurasian Curlew *Numenius arquata* nests on lowland grass-heath. *Bird Study* 66: 471–83. https://doi.org/10.1080/00063657.2020.1725421

34 (a) RSPB (undated) Curlew recovery programme. www.rspb.org.uk/our-work/conservation/projects/curlew-recovery-programme. (b) Grant, M.C., Orsman, C., Easton, J. *et al.* (1999) Breeding success and causes of breeding failure of curlew *Numenius arquata* in Northern Ireland. *Journal of Applied Ecology* 36: 59–74.

35 Macdonald, B. (2019) *Rebirding: Rewilding Britain and its Birds*. Pelagic Publishing, Exeter.

36 Database of Island Invasive Species Eradications. http://diise.islandconservation.org

37 RSPB (undated) Gough Island restoration programme. www.rspb.org.uk/our-work/conservation/projects/gough-island-restoration-programme

38 Thomas, S. and Varnham, K. (2016) Island Biosecurity Manual: Seabird Island Restoration Project. RSPB. https://ww2.rspb.org.uk/our-work/conservation/

shiantisles/work/downloads/hyperlinks/RSPB_Shiants%20LIFE_Biosecurity
%20Manual.pdf
39 Henderson, L. (2009) Progress of the UK Ruddy Duck eradication programme. *British Birds* 102: 680–90.

Chapter 8

1 Lombard, M. and Kyriacou, K. (2020) Hunter-gatherer women. *Oxford Research Encyclopedia of Anthropology.* https://doi.org/10.1093/acrefore/9780190854584.013.105
2 For example, the charitable objects of the Game and Wildlife Conservation Trust: www.gwct.org.uk.
3 Hunting Act 2004. www.legislation.gov.uk/ukpga/2004/37/contents
4 Game and Wildlife Conservation Trust (undated) To bit or not to bit? www.gwct.org.uk/game/research/disease-and-welfare/to-bit-or-not-to-bit
5 Avery, M. (2015) *Inglorious: Conflict in the Uplands.* Bloomsbury, London.
6 Robertson, P.A., Mill, A.C., Rushton, S.P. *et al.* (2017) Pheasant release in Great Britain: long-term and large-scale changes in the survival of a managed bird. *European Journal of Wildlife Research* 63: 100. https://doi.org/10.1007/s10344-017-1157-7
7 There are numerous internet and newspaper reports of pheasant carcasses being dumped presumably because they are surplus to market requirements. This is a typical report: https://markavery.info/2017/01/16/piles-dead-pheasants.
8 In 1985 an estimated 500,000 mallard were released for shooting. No more recent figures are available. Harradine, J. (1985) Duck shooting in the United Kingdom. *Wildfowl* 36: 81–94.
9 British Association for Shooting and Conservation (2019) *Code of Good Shooting Practice.* https://basc.org.uk/codes-of-practice
10 Wild Animal Welfare Committee (2019) The killing and taking of birds under the Wildlife and Countryside Act 1981: a review of the animal welfare implications. https://static1.squarespace.com/static/5edf4fd72d25275e3acc8c4a/t/5f4f9b861c1cad15afc109d5/1599052681811/GLs_-Shooting_and_Trapping_evidence_review.pdf
11 British Deer Society (2019) Deer Stalking Certificate. www.bds.org.uk/index.php/training/dsc1
12 (a) Bateson, P. and Bradshaw, E.L. (1999) How often do stalkers wound red deer? *Deer* 11: 180–181. (b) Bradshaw, E.L. and Bateson, P. (2000) Welfare implications of culling red deer (*Cervus elaphus*). *Animal Welfare* 9: 3–24. (c) Urquhart, K.A. and McKendrick, I.L. (2003) Survey of permanent wound tracts in the carcasses of culled wild red deer in Scotland. *Veterinary Record* 152: 497–501. (d) Aebischer, N., Wheatley, C. and Rose, H. (2014) Factors associated with shooting accuracy and wounding rate of four managed wild deer species in the UK, based on anonymous field records from deer stalkers. *PLOS One* 9(10): e109698. https://doi.org/10.1371/journal.pone.0109698
13 (a) Sneddon, L.U. (2003) The evidence for pain in fish: the use of morphine as an analgesic. *Applied Animal Behaviour Science* 83: 153–62. (b) Sneddon, L.U. (2009) Perception in fish: indicators and endpoints. *ILAR Journal* 50: 338–42. https://doi.org/10.1093/ilar.50.4.338

14 PWC (2018) The contribution of thoroughbred breeding to the UK economy and factors impacting the industry's supply chain. Report to the Thoroughbred Breeders' Association. www.thetba.co.uk/wp-content/uploads/2018/09/TBA-Economic -Impact-Study-2018.pdf

15 League Against Cruel Sports (undated) Horse racing. www.league.org.uk/horse-racing

16 League Against Cruel Sports (2021) Is horse racing cruel? www.league.org.uk/horse -racing

17 Tong, L., Stewart. M., Johnson, I. *et al.* (2020) A comparative neuro-histological assessment of gluteal skin thickness and cutaneous nociceptor distribution in horses and humans. *Animals* 10: 2094. https://doi.org/10.3390/ani10112094

18 Thompson, K., McManus, P., Stansall, D., Wilson, B.J. and McGreevy, P.D. (2020) Is whip use important to Thoroughbred racing integrity? What stewards' reports reveal about fairness to punters, jockeys and horses. *Animals* 10: 1985. https://doi .org/10.3390/ani10111985

19 Wilkins, S. (2019) Whip welts on Melbourne Cup winner (2019) *Horses and People.* https://horsesandpeople.com.au/whip-welts-on-melbourne-cup-winner

20 British Horseracing Authority (undated) Racing Excellence Series. www.british-horseracing.com/regulation/racing-excellence-series

21 (a) Article 8 of The Welfare of Animals at Markets Order 1990 (as amended). www .legislation.gov.uk/uksi/1990/2628/contents/made. (b) Schedule 1 of the Welfare of Animals at the Time of Killing (England) Regulations 2015. https://www.legislation .gov.uk/uksi/2015/1782/contents.

22 Sykes, B., Hewetson, M., Hepburn, R., Luthersson, N. and Tamzali, Y. (2015) European College of Equine Internal Medicine consensus statement: equine gastric ulcer syndrome in adult horses. *Journal of Veterinary Internal Medicine* 29: 1288–99. https://doi.org/10.1111/jvim.13578

23 Royal Society for the Prevention Cruelty to Animals (undated) When the going gets tough on racehorse welfare. www.rspca.org.uk/-/2019_04_04_when_ the_going_gets_tough_on_racehorse_welfare

24 British Horseracing Authority (undated) Making horseracing safer. www.british-horseracing.com/regulation/making-horseracing-safer

25 (a) Dyson, P.K., Jackson, B.F., Pfeiffer, D.U. and Price, J.S. (2008), Days lost from training by two- and three-year-old Thoroughbred horses: a survey of seven UK training yards. *Equine Veterinary Journal* 40: 650–7. https://doi.org/10.2746 /042516408X363242. (b) Wilsher, S., Allen, W.R. and Wood, J.L.N. (2006) Factors associated with failure of Thoroughbred horses to train and race. *Equine Veterinary Journal* 38: 113–18. https://doi.org/10.2746/042516406776563305.

26 Deloitte (2014) Economic impact of the British Greyhound racing industry. A report for the Greyhound Board of Great Britain. www.scribd.com/document/211479827/ The-Economic-Impact-of-the-British-Greyhound-Racing-industry-2014

27 The Welfare of Racing Greyhound Regulations (2010) www.legislation.gov.uk/uksi /2010/543/contents/made.

28 House of Commons Environment, Food and Rural Affairs Committee. Greyhound welfare: Second Report of Session 2015–16.

29 Dogs Trust (2015) The Greyhound Industry: don't bet on fair treatment. www .dogstrust.org.uk/latest/issues-campaigns/greyhounds/the-greyhound-industry-dont -bet-on-fair-treatment

Chapter 9

1 Value of the pet care market in the United Kingdom (UK) in 2020, by category. www.statista.com/statistics/463662/pet-care-market-value-in-the-united-kingdom-by-category

2 Pet Food Manufacturers' Association. Statistics. www.pfma.org.uk/statistics

3 YouGov (2020) Including yourself, do you know anybody who has acquired a new pet during lockdown? https://yougov.co.uk/topics/health/survey-results/daily/2020/10/20/2cdec/3

4 McNicholas, J., Gilbey, A., Rennie, A. *et al.* (2005) Pet ownership and human health: a brief review of evidence and issues. *BMJ* 331: 1252–4. https://doi.org/10.1136/bmj.331.7527.1252

5 Bradshaw, J. (2011) *In Defence of Dogs: Why Dogs Need Our Understanding.* Penguin, London.

6 British Veterinary Association (undated) Brachycephalic dogs. www.bva.co.uk/take-action/our-policies/brachycephalic-dogs

7 Lewis, T. (2018) Here's what popular dog breeds looked like before and after 100 years of breeding. *ScienceAlert.* www.sciencealert.com/what-popular-dog-breeds-looked-like-before-and-after-100-years-of-breeding

8 Farrell, L.L., Schoenebeck, J.J., Wiener, P., Clements, D.N. and Summers, K.M. (2015) The challenges of pedigree dog health: approaches to combating inherited disease. *Canine Genetics and Epidemiology* 2: 3. https://doi.org/10.1186/s40575-015-0014-9

9 Animal Welfare (Licensing of Activities Involving Animals) (England) Regulations 2018. Schedule 6.6(5). www.legislation.gov.uk/uksi/2018/486/schedule/6/made

10 Universities Federation for Animal Welfare (undated) Genetic welfare problems of companion animals: Scottish Fold. www.ufaw.org.uk/cats/scottish-fold-osteochondrodysplasia

11 Universities Federation for Animal Welfare (undated) Genetic welfare problems of companion animals: Manx. www.ufaw.org.uk/cats/manx-manx-syndrome

12 Diesel, G., Pfeiffer, D., Crispin, S. and Brodbelt, D. (2010) Risk factors for tail injuries in dogs in Great Britain. *Veterinary Record* 166: 812–17. https://doi.org/10.1136/vr.b4880

13 British Veterinary Association (2019) Tail docking of dogs: policy statement. www.bva.co.uk/media/1168/tail-docking-rebranded-may-2019.pdf

14 Pegram, C., Raffan, E., White, E. *et al.* (2021) Frequency, breed predisposition and demographic risk factors for overweight status in dogs in the UK. *Journal of Small Animal Practice* 62: 521–30. https://doi.org/10.1111/jsap.13325

15 Spencer, S., Decuypere, E., Aerts, S. and De Tavernier, J. (2006) History and ethics of keeping pets: comparison with farm animals. *Journal of Agricultural and Environmental Ethics* 19: 17–25. https://doi.org/10.1007/s10806-005-4379-8

Chapter 10

1 Speaking of Research. https://speakingofresearch.com/about

2 Cruelty Free International. https://crueltyfreeinternational.org

3 Animals (Scientific Procedures) Act 1986. www.legislation.gov.uk/ukpga/1986/14/contents

4 Directive 2010/63/EU of the European Parliament and of the Council of 22 September 2010 on the protection of animals used for scientific purposes. https://eur-lex.europa.eu/eli/dir/2010/63/2019-06-26

5 Royal Society for the Prevention of Cruelty to Animals and Laboratory Animal Science Association (2015) *Guiding Principles on Good Practice for Animal Welfare and Ethical Review Bodies*. A report by the RSPCA Research Animals Department and LASA Education, Training and Ethics Section (ed. M. Jennings). www.lasa.co.uk/PDF/AWERB_Guiding_Principles_2015_final.pdf

6 Animals in Science Regulation Unit (2015) The harm–benefit analysis process: new project licence applications. Advice Note 05/2015. https://assets.publishing.service.gov.uk/government/uploads/system/uploads/attachment_data/file/487914/Harm_Benefit_Analysis__2_.pdf

7 'Procedures which are performed entirely under general anaesthesia from which the animal shall not recover consciousness shall be classified as "non-recovery".' Home Office (2014) Guidance on the Operation of the Animals (Scientific Procedures) Act 1986, Appendix G. https://assets.publishing.service.gov.uk/government/uploads/system/uploads/attachment_data/file/662364/Guidance_on_the_Operation_of_ASPA.pdf

8 National Statistics (2021) Annual statistics of scientific procedures on living animals, Great Britain 2020. https://assets.publishing.service.gov.uk/government/uploads/system/uploads/attachment_data/file/1002895/annual-statistics-scientific-procedures-living-animals-2020.pdf

9 Russell, W.M.S. and Burch, R.L. (1959) *The Principles of Humane Experimental Technique*. Methuen, London. A digital version of the book may be accessed for free on the website of Johns Hopkins University's Center for Alternatives to Animal Testing: https://caat.jhsph.edu/principles/the-principles-of-humane-experimental-technique.

10 Animals (Scientific Procedures) Act (Amendment) Order 1993. www.legislation.gov.uk/uksi/1993/2103

11 Animals (Scientific Procedures) Act 1986 Amendment Regulations 2012. www.legislation.gov.uk/uksi/2012/3039

12 Home Office (2014) Guidance on the Operation of the Animals (Scientific Procedures) Act 1986.

13 National Statistics (2021) Annual statistics of scientific procedures on living animals, Great Britain, 2020. www.gov.uk/government/statistics/statistics-of-scientific-procedures-on-living-animals-great-britain-2020/annual-statistics-of-scientific-procedures-on-living-animals-great-britain-2020

14 Home Office (2014) Guidance on the Operation of the Animals (Scientific Procedures) Act 1986.

15 Hansard (2008) Animal experiments. Volume 474. https://hansard.parliament.uk/commons/2008-03-25/debates/08032525000061/AnimalExperiments

16 Directive 2010/63/EU of the European Parliament and of the Council of 22 September 2010 on the protection of animals used for scientific purposes. https://eur-lex.europa.eu/legal-content/EN/TXT/?uri=CELEX:32010L0063

17 Jha, A. (2006) Questions raised over ban on research using great apes. *The Guardian*, 3 June 2006 www.theguardian.com/science/2006/jun/03/animalrights.research.

18 The Animals in Science Committee is an advisory non-departmental public body that provides independent advice to the Home Office. www.gov.uk/government/ organisations/animals-in-science-committee/about

19 Animal Procedures Committee (2013) Review of the assessment of cumulative severity and lifetime experience in non-human primates used in neuroscience research. https://assets.publishing.service.gov.uk/government/uploads/system/ uploads/attachment_data/file/261687/cs_nhp_review_FINAL_2013_corrected.pdf

20 Animals in Science Committee (2014) Cumulative severity review: response by Animals in Science Committee. www.gov.uk/government/publications/cumulative -severity-review-response-by-animals-in-science-committee

21 Pycroft, L., Stein, J. and Aziz, T. (2018) Deep brain stimulation: an overview of history, methods, and future developments. *Brain and Neuroscience Advances* 2: 2398212818816017. https://doi.org/10.1177/2398212818816017

22 Research Excellence Framework (2014) UOA05-23: a new form of deep brain stimulation alleviates severe 'freezing' and loss of balance in advanced Parkinson's disease. https://impact.ref.ac.uk/casestudies/CaseStudy.aspx?Id=17518

23 Aguilera, B., Perez Gomez, J. and DeGrazia, D. (2021) Should biomedical research with great apes be restricted? A systematic review of reasons. *BMC Medical Ethics* 22: 15. https://doi.org/10.1186/s12910-021-00580-z

24 Huh, D., Matthews, B.D., Mammoto, A. *et al.* (2010) Reconstituting organ-level lung functions on a chip. *Science* 328: 1662–8. https://doi.org/10.1126/science .1188302

25 Loring, M.D., Thomson, E.E. and Naumann, E.A. (2020) Whole-brain interactions underlying zebrafish behavior. *Current Opinion in Neurobiology* 65: 88–99. https://doi.org/10.1016/j.conb.2020.09.011

26 Stewart, A.M., Braubach, O., Spitsbergen, J., Gerlai, R., and Kalueff, A.V. (2014) Zebrafish models for translational neuroscience research: from tank to bedside. *Trends in Neurosciences* 37: 264–78. https://doi.org/10.1016/j.tins.2014.02.011.

Chapter 11

1 Defra (2018) Health and harmony: the future for food, farming and the environment in a Green Brexit: summary of responses https://assets.publishing .service.gov.uk/government/uploads/system/uploads/attachment_data/file/741461/ future-farming-consult-sum-resp.pdf.

2 Defra (2021) Labelling for animal welfare: call for evidence. www.gov.uk/ government/consultations/labelling-for-animal-welfare-call-for-evidence

3 Government Food Strategy (2022) https://www.gov.uk/government/publications/ government-food-strategy/government-food-strategy#executive-summary

4 PETA (2018) Victory! Companies ban badger-hair brushes after shocking PETA exposé. www.peta.org/blog/companies-ban-badger-hair

Chapter 12

1 When the Treaty of Lisbon came into force in 2009 it amended the Treaty on the Functioning of the European Union and introduced the recognition that

animals are sentient beings. Article 13 of Title II states that: *'In formulating and implementing the Union's agriculture, fisheries, transport, internal market, research and technological development and space policies, the Union and the Member States shall, since animals are sentient beings, pay full regard to the welfare requirements of animals, while respecting the legislative or administrative provisions and customs of the EU countries relating in particular to religious rites, cultural traditions and regional heritage.'*

2 The Animal Welfare (Sentience) Act 2022 https://www.legislation.gov.uk/ukpga/2022/22/enacted

3 Harper, M. (2020) The conservationist's dilemma. Blog post, 7 September 2020. https://community.rspb.org.uk/ourwork/b/martinharper/posts/the-conservationist-s-dilemma-an-update-on-the-science-policy-and-practice-of-the-impact-of-predators-on-wild-birds-7

4 Mosbacher, M. (2022) The rise and fall of Britain's fur trade. *The Spectator*, 29 January 2022 www.spectator.co.uk/article/the-rise-and-fall-of-britains-fur-trade.

5 The Better Chicken Commitment. https://betterchickencommitment.com/en-GB/policy

6 Crustacean Compassion. www.crustaceancompassion.org.uk

7 Wild Justice. https://wildjustice.org.uk

8 Fur Farming (Prohibition) Act 2000. www.legislation.gov.uk/ukpga/2000/33/contents

9 Wild Animals in Circuses Act 2019. www.legislation.gov.uk/ukpga/2019/24/introduction

10 Dubois, S., Fenwick, N., Ryan, E.A. *et al.* (2017) International consensus principles for ethical wildlife control. *Conservation Biology* 31: 753–60. https://doi.org/10.1111/cobi.12896

11 (a) Simmons, A. (2020) Killing badgers to control bTB is unethical. *Veterinary Record* 186: 357–8. https://doi.org/10.1136/vr.m1131. (b) British Veterinary Association (2020) BVA policy position on the control and eradication of bovine TB. www.bva.co.uk/media/3629/bva-policy-position-on-the-control-and-eradication-of-bovine-tb.pdf

12 Scientific Advisory Board on Agricultural Policy (2015) Pathways to a socially accepted livestock husbandry in Germany. Executive summary and synthesis report. Berlin. www.bmel.de/SharedDocs/Downloads/EN/_Ministry/ScientificAdvisoryBoard-Pathways.pdf?__blob=publicationFile&v=2

13 McCulloch, S.P. and Reiss, M.J (2017) The development of an animal welfare impact assessment (AWIA) tool and its application to bovine tuberculosis and badger control in England. *Journal of Agricultural and Environmental Ethics* 30: 485–510. https://doi.org/10.1007/s10806-017-9684-5

14 McCulloch, S.P. and Reiss, M.J. (2018) A proposal for a UK Ethics Council for Animal Policy: the case for putting ethics back into policy making. *Animals* 8: 88. https://doi.org/10.3390/ani8060088

15 Home Office (2014) Guidance on the Operation of the Animals (Scientific Procedures) Act 1986. https://assets.publishing.service.gov.uk/government/uploads/system/uploads/attachment_data/file/662364/Guidance_on_the_Operation_of_ASPA.pdf

16 University of Leicester (undated) The role of lay members in the AWERB. https://le
    .ac.uk/dbs/animal-welfare/awerb/role-of-lay-members-in-the-awerb

17 Ipsos MORI (2013) The public's views on openness and transparency in animal
    research. www.ipsos.com/ipsos-mori/en-uk/publics-view-openness-and-transparency
    -animal-research

18 Cressey, D. (2014) UK institutions sign up to animal-research openness. *Nature*.
    https://doi.org/10.1038/nature.2014.15222

19 Williams, A.J. and Hobson, H. (2021) *Concordat on Openness on Animal Research
    in the UK: Annual Report 2020.* Understanding Animal Research, London. https://
    concordatopenness.org.uk/wp-content/uploads/2021/01/Concordat-Report-2020
    .pdf

Glossary and Abbreviations

Animal rights	The philosophy according to which some, or all, animals are entitled to the possession of their own existence and that their most basic interests – such as the need to avoid suffering – should be afforded the same consideration as similar interests of human beings.
Anticoagulant rodenticide	A substance that, when incorporated into a bait, can be used to poison rodents by interfering with the animal's blood-clotting mechanism.
Appetitive behaviour	An activity that increases the likelihood of satisfying a specific need, such as restless searching for food by a hungry predator.
Animal welfare	The wellbeing of non-human animals. Frequently this is codified in legislation that sets out the duties of animal carers.
ASC	Animals in Science Committee.
AI	Avian influenza or bird flu.
AIHTS	Agreement on International Humane Trapping Standards.
AWERB	Animal Welfare and Ethics Review Body.
ASPA	The Animals (Scientific Procedures) Act 1986.
AWC	Animal Welfare Committee – *see* FAWC.
Behavioural needs	Part of the behavioural repertoire of a particular species that, if it is unable to be expressed because of confinement or other constraint, is believed to be distressing for the individual. Many of these are simple and frequently expressed behaviours such as foraging and hunting for food, and nest building.
BHA	British Horseracing Authority.
Break-back trap	*see* Snap trap.
Broiler chicken	A chicken that is bred and raised specifically for meat production.
BSE	Bovine spongiform encephalopathy.
BTO	British Trust for Ornithology.

238

BVA
British Veterinary Association.

CAK
Controlled atmosphere killing.

Capture-mark-recapture
A technique used to estimate the size of a population where it is impossible or impractical to count every individual. The basic idea is that you capture a small number of individuals, put a harmless, often temporary mark on them, and release them. At a later date, you catch another small group, and record how many have a mark. The proportion of marked animals in the second sample can be used to calculate the size of the population.

Carnivore
An animal whose food and energy requirements derive solely from animal material.

Castration
The surgical removal of the testicles from the male mammal. The main reasons for castration in farmed animals are to limit aggressive behaviour, to prevent boar and bull taint (circulating sex hormones can affect the taste of the meat), and to prevent unplanned breeding.

Cephalopod
Any mollusc of the class Cephalopoda, having tentacles attached to the head, including the cuttlefish, squid and octopus.

CIWF
Compassion in World Farming.

Consummatory behaviour
Behaviour that occurs in response to a stimulus and that achieves the satisfaction of a specific drive, such as the eating of captured prey by a hungry predator.

Corvids, Corvidae
A cosmopolitan family of passerine birds that contains the crows, ravens, rooks, jackdaws, jays, magpies, choughs, and nutcrackers.

Decapod
A crustacean such as shrimp, lobsters, and crabs with five pairs of thoracic appendages or legs (hence decapod – ten legs), one or more of which are modified into pincers.

Deontology
A theory that suggests actions are good or bad according to a clear set of rules. The word comes from the Greek word *deon*, meaning duty. Actions that obey these rules are ethical, while actions that do not, are not.

Defra
Department for Environment, Food and Rural Affairs.

Dehorning and disbudding
Dehorning and disbudding are forms of mutilation. Dehorning is the surgical removal of the horn from an adult bovine (or goat). Disbudding is the procedure applied to a young animal where the tissue from which the horn will grow is destroyed, generally by cautery (the application of intense, localised heat

239

to destroy tissue or control haemorrhage). In Britain, the procedure must always be done under local or general anaesthesia.

EFRACom House of Commons Environment, Food and Rural Affairs Select Committee.

Ethics The set of moral principles that guide a person's behaviour. These morals are shaped by social norms, cultural practices and religious influences. Ethics reflect beliefs about what is right, what is wrong, what is just, what is unjust, what is good, and what is bad in terms of human behaviour.

FAO Food and Agriculture Organization of the United Nations.

FAWC The Farm Animal Welfare Committee was an independent advisory body established by the UK government in 2011. It replaced the Farm Animal Welfare Council, which was established in 1979. The Council published its final report before its closure and replacement in 2019 by the Animal Welfare Committee (AWC). AWC has an extended remit and includes farmed, companion and wild animals kept by people.

Feed conversion ratio A measure of the efficiency by which feed provided to an animal in converted to body mass. The lower the ratio the more efficient the animal. FCR is calculated by dividing the weight of feed eaten daily by the amount of live weight gained daily. For instance, if a pig ate 1 kg of feed a day and put on 0.5 kg in weight its FCR for this period would be 2. It is never quite as simple as that since FCR varies with diet, age, environment, health status, etc.

In general, over its lifespan between hatching and slaughter a broiler chicken has an FCR of 1.9. This compares with cattle (FCR = 8), pigs (3.9) and salmon (1.3).

GPS Global Positioning System.

Great apes For the purposes of this book, great apes are gorillas, chimpanzees, bonobos and orang-utans. In the United Kingdom, a ban on their use in research has been in place since 1996.

Harem A group of female animals, generally herbivores, controlled by a male of the same species to secure exclusive mating.

Herbivore An animal anatomically and physiologically adapted to eating plant material.

IEP	Independent Expert Panel on Badger Culling Pilots.
in vitro	Latin for 'in glass'. It refers to medical procedures, tests and experiments performed outside a living organism. An *in vitro* study occurs in a controlled environment, such as a test tube or petri dish.
in vivo	Latin for 'in the living'. It refers to research that is performed on a whole, living organism.
IUCN	International Union for the Conservation of Nature.
IWC	International Whaling Commission.
Laying hen	A female, grown chicken that is bred and kept primarily for laying eggs for human consumption.
LACS	League Against Cruel Sports.
Lairage	A place, generally in the form of pens, where farmed animals may be held prior to slaughter.
Monogastric animal	A mammal with a single glandular stomach, including humans, poultry, pigs, horses, rabbits, dogs and cats. Most monogastrics are generally unable to digest much cellulose food materials such as grasses.
Mutilation	The surgical removal of part of an animal's body, such as tail docking, dehorning and castration.
Mulesing	A surgical procedure during which the skin around the breech and tail area of Merino sheep is removed. It is usually carried out on young sheep before they reach six months of age. Once the wound has healed the skin becomes tight around the tail area and very little wool grows there, thereby reducing the risk of flystrike. The procedure is widely used in Australia but prohibited in Britain.
NE	Natural England.
Neuroscience	The science concerned with the study of the structure and function of the nervous system.
NGO	Non-governmental organisation.
OECD	Organisation for Economic Co-operation and Development.
Omnivore	An animal that has the ability to eat and survive on both plant and animal matter.
Operant conditioning	A method of learning where the consequences of a response determine the probability of it being repeated. Through operant conditioning, behaviour which is reinforced (rewarded) will likely be repeated, and behaviour which is punished will occur less frequently.
Osteoporosis	Osteoporosis in laying hens is a condition that involves the progressive loss of structural bone during the laying period. This bone loss results in increased

bone fragility and susceptibility to fracture, with fracture incidences of up to 30% over the laying period.

Primate — Any mammal of the group that includes lemurs, lorises, tarsiers, monkeys, apes, and humans.

Reintroduction — Restoring a species from a place from where it has been lost.

Ruminant — Any mammal of the suborder Ruminantia (order Artiodactyla), which includes deer, cattle, antelopes, sheep and goats.

RSPCA — Royal Society for the Prevention of Cruelty to Animals.

RSPB — Royal Society for the Protection of Birds.

SNH — Scottish Natural Heritage, renamed NatureScot in 2020.

SBS — SongBird Survival.

Sleep apnoea — Sleep apnoea is common in obese dogs and flat-faced breeds such as English Bulldogs, Boston Terriers and Pugs. Excessive internal fat or an abnormal respiratory anatomy can temporarily collapse or narrow the airway, forcing the dog to wake. Badly affected dogs will sleep poorly and fitfully.

Snap trap — A type of lethal spring trap for killing mice and rats. Unlike spring traps for stoats and other small mammals, these type of traps are exempt from any statutory approval requirement.

Spring trap — Any type of lethal spring-operated trap used to catch and kill small to medium-sized mammals. Spring traps include snap or break-back traps and the now prohibited pole traps, previously used to catch and maim birds of prey.

Stereotypical behaviour — A behaviour pattern that is repetitive and unvarying. Stereotypical behaviours, or stereotypies, are often performed by captive animals; they include pacing by captive polar bears, bar chewing by sows and crib biting by horses. Stereotypies are abnormal, both in the sense of having no apparent function and often also in the sense of being seemingly absent in the 'normal', that is, the healthy and free-living, animal. They are of concern they because they are often associated with poor welfare.

Tail docking — A mutilation where part or all of the tail of animal is amputated. Reasons for tail docking vary from species to species. For example, in sheep it is to help

	keep the rear of the animal clean. In dogs, it is argued that tail docking protects the tail from injury.
telos	According to the Greek philosopher Aristotle, the *telos* of an animal is 'what it was made for'. In other words, it can be argued that a dog was 'made' to be a pack animal and a cat 'made' to be a hunter.
The 3Rs	A set of principles to govern aspects of the use of animals in science. The 3Rs are replace, reduce and refine.
Translocation	The human-mediated movement of one or more wild animals from one place to another for any reason.
Utilitarianism	An ethical theory that determines right from wrong by focusing on outcomes. Utilitarianism holds that the most ethical choice is the one that will produce the greatest good for the greatest number.
Vivisection	The practice of performing operations on live animals for the purpose of experimentation or scientific research. This term tends to be used only by the opponents of the use of animals in science.
WWT	Wildfowl and Wetlands Trust.
Xenotransplantation	The transplantation of living cells, tissues or organs from one species to another.
Zero-grazing	A type of dairy farming in which the cattle are fed with cut grass, permanently housed and rarely if ever have access to pasture.
Zoonosis, zoonotic	A disease transmissible between vertebrate animals and humans.

Further Reading

Avery, M. (2015) *Inglorious: Conflict in the Uplands.* Bloomsbury, London.

Brambell, R. (1965) *Report of the Technical Committee to Enquire Into the Welfare of Animals Kept Under Intensive Livestock Husbandry Systems*, Cmd. (Great Britain. Parliament), HM Stationery Office.

de Waal, F. (2016) *Are We Smart Enough to Know How Smart Animals Are?* Granta, London.

Harrison, R. (2013) *Animal Machines: The New Factory Farming Industry*, revised edition. CABI Publishing, Wallingford.

Macdonald, B. (2019). *Rebirding: Rewilding Britain and its Birds.* Pelagic Publishing, Exeter.

Singer, P. (1975) *Animal Liberation: Toward an End to Man's Inhumanity to Animals.* Cape, London.

Singer, P. (2011) *Practical Ethics*, 3rd edition. Cambridge University Press, Cambridge.

Tree, I. (2018) *Wilding: The Return of Nature to a British Farm.* Picador, London.

Acknowledgements

Writing a book, as I found out, is a solitary task. Perhaps the enforced isolation of the Covid pandemic was the ideal time to start writing seriously. Despite the isolation, however, I have enjoyed support from colleagues and friends throughout – both remotely and, as things started to relax somewhat, face to face. I am grateful to them all.

This book would never have been started – let alone finished – if Nigel Massen of Pelagic Publishing had not contacted me, completely unexpectedly, to ask me if I was interested in writing a book. I owe him my thanks for making that initial approach, and for his unstinting support and advice throughout. Having my work edited by professionals was a completely new experience. Thanks to Hugh Brazier and David Hawkins the entire process was satisfying, productive and largely painless.

Stuart Reeves and Ian Davidson deserve special thanks. Not only did they read and comment on several chapters, but over the last few years they have endured hours of me practising the arguments set out in this book. They were, on our numerous birding trips together, little more than a captive audience. Whether we were walking the forests of Finland or scouring the deserts of Morocco, both remained receptive and largely good-humoured despite getting their ears relentlessly bent about the killing of badgers or the welfare of poultry.

Ian Carter, author and ornithologist, commented on the entire text, and his wise head has helped to keep it relevant and accurate.

I am grateful to Amanda Barrett who, on commenting on the first complete draft, provided a vegan perspective.

James Lowen, author, naturalist and friend, has throughout the process offered useful advice about the process of writing and publishing – not to be ignored, whether given over the phone or while watching waders together in Norfolk.

Various friends and colleagues commented on early drafts of the text. For the most part they were perhaps not harsh enough in their criticism, but I am grateful nonetheless. Veterinarians Pete Goddard, Liz Mullineaux and Anthony Wilkinson were particularly helpful in ensuring the text was technically sound. Huw Golledge, a colleague and a friend, provided sound advice on the treatment of animals used in science. Advice from Matt Cross and Paul Tout on the regulation and use of firearms and shotguns is much appreciated.

Damon Bridge and Alison Morgan, stalwarts of the Great Crane Project in Somerset, introduced me to the joy and excitement of catching and ringing baby cranes. Damon Bridge was a mine of information on the crane reintroduction project and the associated predator mitigation measures.

I am grateful to Sandra Baker of Oxford University for her kind advice on the various mostly inhumane methods of killing rodents.

In 2019, Mark Avery provided me with a platform on his blog where I was able to test, for the first time in print, some of the book's themes.

Finally, my wife Liz and my children, Lucy and Robin. They have their own lives to lead and we wisely respected interpersonal distances while I toiled in the office. But once they got over their justifiable scepticism ('The old man is on another crazy project'), they took a keen interest in progress and provided much-needed encouragement as well as plenty of afternoon tea.

Index

References to endnotes are indicated by the letter n following the page number.

abuse of animals   2, 6, 11, 16, 83, 146, 193
Agreement on International Humane Trapping Standards (AIHTS)   112, 226n17
alphachloralose   122
angling   151–52
Animal (Scientific Procedures) Act 1986   38, 169, 214n21
animal diet   21, 35, 64, 76, 80, 81, 85, 90, 98, 100, 153, 154, 162, 165
animal experimentation   10, 166–80, 209
animal exploitation   1, 2, 19–20, 190; consumers making decisions about 17–18; decisions about 187; defending of 6; development of values 7–11; of dogs and cats 158–59; ethics of 204, 206–7; moral imperative 5–7; for research 167–68; scale of 4–5; setting limit to tolerate 197; of wildlife 102–4
'animal exploitation establishment'   3
Animal Health Act 1981, Sections 40 and 41 of   215n2
Animal Health and Welfare (Scotland) Act 2006   106
Animal Liberation (Singer)   15–16
Animal Machines (Harrison)   31, 46, 92
animal production   1, 10, 46, 58
animal products: consumption of 45–46; other than food   186
animal protection law/legislation   31–33
animal rights   15–16
animal rights theory   192–93
Animal Sentience Committee   207
animal welfare   13, 19; dairy herds threaten 70–71; efforts to ensure

compliance 38; need for transparency 129–31; not same as animal rights 192–93; public pressure for 18; risks to 190
Animal Welfare Act (2006)   26, 106, 122, 192, 201, 225n3
Animal Welfare (Sentience) Act 2022   14, 194, 216n22
Animal Welfare and Ethical Review Body (AWERB)   19, 167, 170, 176, 180, 208, 209, 214n21
Animal Welfare Committee (AWC)   170
Animal Welfare Impact Assessment (AWIA)   207
animal welfare law/legislation   18, 31–32, 106, 204
animal welfare science   42
animals: arbitrary groupings perpetuate inconsistent treatment of 191–92; in sport 144, 152–56. *See entries on individual species*
Animals in Science Committee (ASC)   175, 176, 235n18
Animals (Scientific Procedures) Act 1986 (ASPA)   38, 169–71, 174, 179
anti-coagulant rodenticides   2, 26, 122, 140
appetitive behaviour   35
Are We Smart Enough to Know How Smart Animals Are? (de Waal)   11
ASC. *See* Animals in Science Committee (ASC)
Atlantic salmon *Salmo salar*   97–100
attitudes: change over time 30–31; to farmed animals 44–46; to wild animals 23

Australia, cattle and sheep in 24
Avery, Mark 203; *Inglorious* 147
AWC. *See* Animal Welfare
    Committee (AWC)
AWERB. *See* Animal Welfare and Ethical
    Review Body (AWERB)
AWIA. *See* Animal Welfare Impact
    Assessment (AWIA)

bacon pigs 82
badger *Meles meles* 186; killing of 107–9,
    124, 204–5
barnacle goose *Branta leucopsis* 131
basic research 10, 172
bats, tagging of 134
battery cage system 91–93, 95
beaver *Castor fiber* 135
beef 182; annual consumption of 61;
    carbon footprint from 188
beef cattle 62–63
behavioural needs 6; of animals 35–36;
    differences are adaptive 36–37
behavioural repertoire 36, 49, 62, 86, 92,
    158, 161, 165
Belize 77–78, 150, 151
Bentham, Jeremy 13
Better Chicken Commitment 89, 90,
    201, 202
BHA. *See* British Horseracing
    Authority (BHA)
biodiversity 188
Biodiversity Intactness Index 135
birds: abundance and diversity of 139;
    ringing, tagging and other marking
    techniques for 131–34; shooting
    of 148–51
Birkhead, Tim: *Bird Sense* 94
Blakemore, Colin 175
bovine spongiform encephalopathy
    (BSE) 28, 57
Brambell, Francis William Rogers 33
breeding 39; of GM laboratory
    animals 172; pigs 81; poultry 40,
    84–86, 89
British and Irish bird ringing scheme 132
British Horseracing Authority
    (BHA) 154
British Trust for Ornithology (BTO) 132,
    133, 192

British Veterinary Association (BVA) 56,
    63, 73, 109, 122, 162
broiler chicken 4, 6, 27–28, 40, 46,
    85–90; protein in diet 98
broodiness 49
brown rat *Rattus norvegicus* 26
BSE. *See* bovine spongiform encepha-
    lopathy (BSE)
BTO. *See* British Trust for
    Ornithology (BTO)
Burch, Rex: *The Principles of Humane
    Experimental Technique* 172
Burns, Robert 166
Buzzard *Buteo buteo*, killing of 136–37;
    recovery of population 136
BVA. *See* British Veterinary Association

controlled atmosphere killing (CAK)
    systems 55–56
Canada goose *Branta canadensis* 141
captive animals: ecology of ancestral
    form 39–41; welfare of 41–42
carbon footprint 188–89
castration 62, 63, 72–73, 79
cats 157; exploitation of 158–59; Manx
    161; mutilations 161–62; obesity 162;
    Scottish Fold 160–61
cattle 61–62; beef cattle 62–63; dairy
    cattle 63–64; health and welfare of
    64–65; high-yielding dairy cows 66,
    70, 75; infertility 65–66; lameness 66;
    mastitis 65; mutilations 67–68
cell culture 10, 178
cephalopods 14, 216n22
chicken meat 27–28, 85
chimpanzees 11
Churchill, Winston 77
citizen 199–200; assembly of 208;
    exerting influence as 200–202
climate change 75, 104, 119, 188
Code of Practice (CoP) 15, 33, 114
*Code of Practice for the Welfare for Laying
    Hens and Pullets* (Defra) 94
Code of Practice for the Welfare of Pigs
    (Defra) 80
Codes of Welfare Practice (Defra) 32
Compassion in World Farming
    (CIWF) 92, 97, 99
compassionate conservation 129–30

computer models 178
conservation 127–28, 141–43; organisa-
    tions exploit wildlife 128–29
conservation NGOs 128–30, 143, 198
consumer 199–200; exerting influence
    as 200
consummatory behaviour 35
controlled shooting 108–9
Cornwall cirl bunting reintroduction
    project 135–36
corvids 113, 118
cranes 127–28; mortality rate of 133
cross-breed dogs 160
Cruelty Free International 169
Cruelty to Animals Act 1876 169
Crustacean Compassion 201, 202
crustaceans 183

dairy calves, health and welfare of 68–70
dairy cattle 63–64
dairy farming 75
Dawkins, Marian Stamp 13, 14
DBS. See deep brain stimulation
de Waal, Frans: Are We Smart Enough to
    Know How Smart Animals Are? 11
decapods 14, 174, 216n22
decision making, wildlife 123–26
deep brain stimulation (DBS) 177
deer 148; shooting of 150–51
Defra 94, 109, 114
dehorning 67, 68
deontology 192
Descartes, René 12, 13, 17
diet: animal 21, 35, 64, 76, 80, 81,
    85, 90, 98, 100, 153, 154, 162, 165;
    human 28, 44, 87
disbudding 67, 68
disease: eggs 91; farmed animals 50; in
    village pigs 78
Diseases of Animals Act 215n2
dogs 25, 36, 157; exploitation of 158–59;
    hunting with 145–46; mutilations
    161–62; obesity 162
Dogs Trust 156
domestication 163
ducks 96–97, 141

egg-laying hens 90–96
eggs 28, 183; mandatory labelling of 184

electrical stunning 56
environmental impacts 188
Ethical Advisory Committee, RSPB 130
ethical wildlife control 205
ethics 9, 19; of animal exploitation
    204, 206–7
EU legislation 88, 92, 96, 170,
    175, 214n10
EU organic standards 96
Eurasian crane Grus grus. See cranes
European Council 194
European mole Talpa europaea 116–17
European Parliament 194
European Union 33, 55, 88, 89, 92,
    93, 168
evidence-based campaigns 201
evidence-based decision making 203–5
exotic animals, as pets 163

familiarity: lack leads to fear and
    intolerance 29–30; may not
    breed contempt, but it drives
    indifference 27–28
farm animal welfare 35–36
Farm Animal Welfare Committee/Council
    (FAWC) 65, 93, 98–100, 125
farmed animals: breeding 40–41; in
    Britain 38; disease 50; experiences and
    attitudes 44–46; husbandry of 46–48;
    killing of 151; mutilations 50–52;
    sheep/deer 37; slaughter 52–56; trans-
    portation 52–54
farmed fish 97–100, 183
farming: of animals in Britain 44, 46;
    impact on natural behaviour 48–49;
    protection of 115–18
farming organisations 198
fear 29–30
Fearnley-Whittingstall, Hugh 145
feed conversion ratio (FCR) 86
fisheries, protection of 118–19
Five Domains model 34, 75
Five Freedoms 32–34, 93
food: killing animals for 146–48;
    labelling of 184–85; storage and
    preparation of 105
food animal production 1
fore-gut fermenters 60
fox hunting 17, 23–24, 42, 202

free-living wild animals 103
free-range systems 95
full-cream milk 63
fur farming 204
furnished cages 92–94

game birds 110–11, 182
gas stunning 55
gastric ulcers 154
geese 22, 101, 113n22, 131, 141, 148
General Licences 106–7, 112, 118,
    119, 149
genetically modified (GM) animals 172
giant dog breeds 159
global heating 188
glue boards 121–22
Gough Island 140
grazing 57–58
grazing animals 59–61
Great Crane Project 127
Greenland white-fronted goose *Anser
    albifrons* 131
grey partridge *Perdix perdix* 111, 138
grey squirrel *Sciurus carolinensis* 30
greyhound racing 155–56
Greylake RSPB 138
ground-nesting bird species 137–38
grouse 111, 144, 147–48

harm–benefit analysis 170–71, 195,
    210, 211
Harrison, Ruth: *Animal Machines* 31,
    46, 92
Hisex Brown 85, 222n20
horses/racehorses 2–4, 8, 23, 33, 60, 63,
    145, 146, 152–56, 169, 170, 174, 192,
    215n2; high-protein diet 153–54
horse meat 23
horticulture 60
House of Commons Environment,
    Food and Rural Affairs Committee
    (EFRACom) 156
Hughes, Ted ('Dehorning') 67
human diet 28, 44, 87
human tissue 178
Humane Trapping Standards Regulations
    2019 226n17
humaneness 107–9, 115; of shooting
    deer 150

human-mediated deformities 159–61
hunting 15, 144; with dogs 145–46; fox
    17, 23–24, 42, 202
Hunting Act 2004 145, 146
husbandry 32, 43; of farmed
    animals 46–48

IEP. *See* Independent Expert Panel
illegal killing 148; of wild mammals 146
inclusive decision making 203–5
inconsistencies in legal
    protection 31, 191–92
Independent Expert Panel (IEP) 108, 109
Individual Licence 107, 119
infertility, adult cow 65–66
Inglorious (Avery) 147
International Association for the Study of
    Pain 12
International Union for the Conservation
    of Nature (IUCN) 27, 135
International Whaling Commission
    (IWC) 25
intolerance 22, 29–30
Ipsos MORI 209
Islay Sustainable Goose Management
    Strategy 131
IUCN. *See* International Union for the
    Conservation of Nature (IUCN)
IWC. *See* International Whaling
    Commission (IWC)

K-selection strategy 39

labelling, for packaged food 184–86
LACS. *See* League Against Cruel
    Sports (LACS)
lamb 58, 72, 73, 98, 183; carbon
    footprint from 188
lameness: adult cows 66; sheep 73–74
League Against Cruel Sports
    (LACS) 24, 153
'lethal scaring' 131
lethal traps 111–12, 121, 195
licence 117; types of 106–7
live-capture traps 112–13, 117, 122, 195
livestock farming 9, 31, 44, 46, 71,
    75, 188
live-trapping: of crows 141; of
    rodents 123

McCulloch, Steven   207
Macdonald, Benedict: *Rebirding*   139
mammalian predators   119
mammals: ringing, tagging and marking
    techniques for   131–34; shooting
    of 148–51
Manx cat   161
marking techniques, for birds and
    mammals   131–34
mastitis, adult cow   65
meat industry   198
Minsmere RSPB   139
mole   116–17
mongrel dogs   160
Munro, Ranald   24
mutilations   50–52; cattle 67–68; pigs
    79–81; sheep 72–73

national dairy herd   64
National Farmers Union's Red
    Tractor   184
National Trust   146
Natural England (NE)   58, 109, 112,
    127, 136
natural weaning   69
NE. *See* Natural England
needle teeth removal, pigs   79–80
neuroscience research   176
non-domesticated animals   106
non-native rodents   140
non-native species, killing and ultimate
    elimination of   141
non-retrieval rate (NRR), in
    shootings   109
non-target species   113
nose ringing, pigs   81
numbers of: animals used in research
    171–75; badgers killed annually 204;
    farmed animals 47, 70–71; 'game' birds
    reared and killed annually 147; rodents
    killed annually 5, 192

obesity in dogs   166–67
Open Farm Sunday   198
organ-on-a-chip   178–79
osteoporosis, in poultry   93, 94

packaged food, labelling for   184–86
Packham, Chris   203

pain   12–14; perception of   24
Parkinson's disease   177
pedigree dogs   159
People for the Ethical Treatment of Animals
    (PETA)   186
peregrine falcon *Falco peregrinus*   137
personal ethical framework   181–83;
    animal products other than food 186;
    labelling for packaged food 184–86;
    putting into practice 187–89
pest control   111, 123–27
pest species, protecting businesses and
    households from   119–23
pet food market   157
pets   157–58; chronic conditions and
    geriatric 163–64; exotic animals as 163;
    human-mediated deformities 159–61;
    mutilations 161–62; obesity 162; repre-
    sents an ethical dilemma 164–65
pheasants   5, 18, 104, 105, 110–11, 136–
    37, 144–45, 147, 148, 149, 197, 201
pigs   40, 76–84; commercial diets of
    76; confinement and barren environ-
    ments 81–83; digestive system of 76;
    knowledge of 77; mutilations 79–81;
    welfare of 83–84
pine marten *Martes martes*   30, 102, 112
pink-footed goose *Anser
    brachyrhynchus*   101
point of lay (POL) hens   91
polecat *Mustela putorius*   102, 106, 112
pork   53, 77, 82, 84, 183
poultry   84–85; broiler chickens 85–90;
    commercial diets of 76; egg-laying hens
    90–96; slaughter of 55; turkeys and
    ducks 96–97
poultry meat   4, 28, 54, 182, 188
predator control: on islands   139–41; to
    protect threatened species   136–39
primates, use in animal research   174–75,
    177
*The Principles of Humane Experimental
    Technique* (Russell and Burch)   172
Protection of Animals Act 1911   31
Protection of Badgers Act 1992   106

racing: of greyhounds   155–56; of
    horses   152–55
racing industry   145

rats 5, 11, 22, 26, 100, 103, 105, 139, 175–77, 192, 210; GM rats 172; poisons 122; protecting businesses and households from 119–23; protecting farms from 115; spring traps for 112
Rebirding (Macdonald) 139
recreation, killing animals for 146–48
red grouse *Lagopus lagopus scotica* 147
red squirrel *Sciurus vulgaris* 30
Red Tractor scheme 17, 214n19
red-legged partridge *Alectoris rufa* 5, 111, 147
Reeves, Stuart 139
regulated spring traps 111–12
regulatory testing 172
reintroductions 135–36, 142
Reiss, Michael 207
research: animals used in 173–77; in disease 10; establishments of 198–99; killing of animals in 192; public engagement in 208–9; types of 170; use of animals in 32
research animals 166–67, 179–80; alternatives to using 178–79; breeding 41; exploitation of 167–68; harm *versus* benefit 170–71; numbers count 171–72; primates 174–75, 177; regulation and controls 168–70; replacing higher animals with lower animals 179; 3Rs 172–73
rifles 148–49
ringing techniques, for birds and mammals 131–34
rodenticides/rat poisons 122
rodents: controlling in home and food premises 140; killing of 102, 122, 123, 140, 141, 197; managing in domestic environments 141; non-native 140; plastic traps 121; problems in house 120; signs of respiratory irritation 118; traps used for 121; use of poisons 2, 5, 21
Royal Commission on Vivisection 169
Royal Society for the Prevention of Cruelty to Animals (RSPCA) 17, 97, 154, 214n19
Royal Society for the Protection of Birds (RSPB) 57, 58, 127, 130, 131, 140, 198; Curlew Recovery Plan 138

r-selection strategy 39
RSPB. *See* Royal Society for the Protection of Birds (RSPB)
RSPCA Assured scheme 17, 184, 185
ruminants 60, 98
Russell, William: *The Principles of Humane Experimental Technique* 172

SBS. *See* SongBird Survival (SBS)
Scottish Animal Welfare Commission 125
Scottish Anti-Vivisection Society 166
Scottish Fold cat 160–61
Scottish Natural Heritage (SNH) 131, 226n19
Second World War 64
selective breeding 89, 97
self-awareness in animals 16
sentience 14–15, 193–96
sheep 54, 59, 71–75; husbandry 58; lameness 73–74; mutilations 72–73
shooting 147–48; of birds and mammals 148–51; non-retrieval rate in 109; organisations 198; protecting birds as quarry for 110–11
shotguns 148–49
Singer, Peter: *Animal Liberation* 15–16
slaughter, farmed animals 52–56
slaughterhouses 24, 52, 53–54, 55, 153, 192, 199
snap traps 121, 123. *See also* spring traps
snares 113–15
Sneddon, L.U. 14
SNH. *See* Scottish Natural Heritage
social groups 49
Soil Association 184–85; organic standards 96
SongBird Survival (SBS) 137
sparrowhawk *Accipiter nisus* 137
sport, animals in 152–56
spring traps 111–12; for killing stoats 192
statutory conservation bodies 130–31, 133
stocking density 47–48, 88, 99, 222n27
stunning methods 55–56
suckler herds 62
suckling 64, 69, 118
suffer: capacity to 12–14; perception of 24

tagging techniques, for birds and
    mammals   131–34
tail docking   50, 72, 73, 80, 84, 162
3Rs, research animals   73, 172–73, 178–80
Tingay, Ruth   203
toy dog breeds   159
translocations   135–36, 142
transparency, need for   129–31
transportation, farmed animals   52–54
traps   120–21; lethal 111–12; live
    capture in 112–13; approval of orders
    across UK administrations 225–26n17;
    snap 121, 123; spring 111–12. *See
    also* snares
Treaty of Lisbon (2009)   194, 235n1
turkeys   96–97

Usable Agricultural Area   60
utilitarianism   10, 16, 17, 168, 180,
    193, 211

values, development of   7–11
Vegan Society   213n1
venison   145, 152, 183
Veterinary Surgeons Act 1966   63
village pigs   78–79

Webster, John   83
welfare labelling   184, 185

Welfare of Animals Act (Northern Ireland)
    2011   106
whales, whaling   25, 27
wild animals, attitudes to   23
wild boar *Sus scrofa*   29, 40
Wild Justice   203
Wild Mammals Protection Act 1996   106
wildfowling   148
wildlife   5, 38; activities affect 104–6;
    breeding 41; in Britain 102; conser-
    vation organisations exploit 128–29;
    decision making and accountability
    123–26; exploitation of 102–4; ignore
    suffering of 41–42; law protect from
    harm 106–7; management problem
    120; objectives of killing 124
Wildlife and Countryside Act
    1981   106, 149
wildlife management,
    influence   142, 202–3
Wildlife Trusts   128
wolf *Canis lupus*   29, 36, 158
Wood-Gush, David   93

YouGov   214n20

zebrafish *Danio rerio*   41, 179
zero-grazing systems   206
Zoological Society of London   136

About the Author

Alick Simmons is a veterinarian and a naturalist. After a 35-year public service career controlling epidemic diseases of livestock, culminating in eight years as the UK's Deputy Chief Veterinary Officer, in 2015 he began conservation volunteering. As well as practical tasks such as surveying waders and catching baby cranes, he advises a number of conservation organisations on animal welfare and ethics. He is chair of the Zoological Society of London's Ethics Committee for Animal Research and sits as an independent member of ethics committees for both the RSPB and National Trust. He is former chair of the Universities Federation for Animal Welfare and the Humane Slaughter Association.

Also available from Pelagic Publishing

Traffication: How the Car Killed the Countryside, Paul F. Donald (coming spring 2023)

Invisible Friends: How Microbes Shape Our Lives and the World Around Us, Jake M. Robinson (coming spring 2023)

Reconnection: Fixing Our Broken Relationship with Nature, Miles Richardson (coming summer 2023)

Reflections: What Wildlife Needs and How to Provide It, Mark Avery (coming autumn 2023)

Low-Carbon Birding, edited by Javier Caletrío

The Hen Harrier's Year, Ian Carter and Dan Powell

Wildlife Photography Fieldcraft: How to Find and Photograph UK Wildlife, Susan Young

Rhythms of Nature: Wildlife and Wild Places Between the Moors, Ian Carter

Ancient Woods, Trees and Forests: Ecology, Conservation and Management, edited by Alper H. Çolak, Simay Kırca and Ian D. Rotherham

Essex Rock: Geology Beneath the Landscape, Ian Mercer and Ros Mercer

Pollinators and Pollination, Jeff Ollerton

The Wryneck: Biology, Behaviour, Conservation and Symbolism of Jynx torquilla, Gerard Gorman

Wild Mull: A Natural History of the Island and its People, Stephen Littlewood and Martin Jones

Challenges in Estuarine and Coastal Science, edited by John Humphreys and Sally Little

A Natural History of Insects in 100 Limericks, Richard A. Jones and Calvin Ure-Jones

Writing Effective Ecological Reports, Mike Dean

pelagicpublishing.com